LASERS, DEATH RAYS, AND THE LONG, STRANGE QUEST FOR THE ULTIMATE WEAPON

LASERS, DEATH RAYS, AND THE LONG, STRANGE QUEST FOR THE ULTIMATE WEAPON

JEFF HECHT

59 John Glenn Drive
Amherst, New York 14228

Published 2019 by Prometheus Books

Inquiries should be addressed to
Prometheus Books
59 John Glenn Drive
Amherst, New York 14228
VOICE: 716–691–0133 • FAX: 716–691–0137
WWW.PROMETHEUSBOOKS.COM

23 22 21 20 19 5 4 3 2 1

Library of Congress Cataloging-in-Publication Data

Names: Hecht, Jeff, author.
Title: Lasers, death rays, and the long, strange quest for the ultimate weapon / by Jeff Hecht.
Description: Amherst, New York : Prometheus Books, [2019] | Includes index.
Identifiers: LCCN 2018035104 (print) | LCCN 2018036682 (ebook) |
ISBN 9781633884618 (ebook) | ISBN 9781633884601 (hardcover)
Subjects: LCSH: Laser weapons—History. | Laser weapons—Design and construction.
Classification: LCC UG486 (ebook) | LCC UG486 .H43 2019 (print) | DDC 623.4/46—dc23
LC record available at https://lccn.loc.gov/2018035104

Printed in the United States of America

CONTENTS

CHAPTER 1

DEATH RAYS: FROM THUNDER GODS TO MAD SCIENTISTS

Ancient myths armed powerful gods with death rays. The long list of thunder gods on *Wikipedia*—the unruly mirror of our culture—covers the globe and history, ranging from Zeus, the king of the Greek gods, to heavy metal musician Gene Simmons, whose trademark song for KISS is "God of Thunder."[1] The ancients didn't understand the nature of lightning, but they knew and feared it. To them, lightning was the first directed-energy weapon, hurled by the gods to smite the evil, the unworthy, or those who had annoyed them. The flash of light came first, and if the strike was close, the bang and an explosive burst of energy followed almost instantly, shattering and igniting trees and killing people.

Lightning follows a zigzag path across the sky. A laser beam is an unerringly straight line through the air. Both are directed energy, but nature directs electricity differently than light. Death rays are modern myths, updated versions of the mythic bolts hurled by mythic ancient gods, born more than a century ago at a time when scientists were puzzling over new discoveries from X-rays to radio waves, inventors were seeking new weapons of war, and storytellers were looking for thrilling new ways to entertain their audiences.

The laser is a newcomer, invented in 1960 by Theodore Maiman, a young physicist who built on Albert Einstein's new physics and the work of others. He had high hopes for uses of his new discovery in research, communications, industry, and medicine when he announced it at a Manhattan press conference. But he was dismayed after he returned home to Los Angeles to find that newspaper headlines were mostly variations on "LA Man Discovers Science Fiction Death Ray."[2] This chapter tells the origins of the death-ray myth that waited for Maiman's new invention.

ARCHIMEDES AND THE BURNING MIRRORS

The oldest and most intriguing tale of using light as a weapon involves Archimedes, a legendary figure often considered the greatest mathematician and engineer of ancient times. He lived in Syracuse, a Greek city in what is now Sicily, at a time when Rome was expanding its power. When the Romans began threatening Syracuse in 215 BCE, Archimedes designed a series of war engines that used catapults, Greek fire, and other mechanisms to fend them off. Also he is said to have orchestrated the use of a large array of mirrors to focus sunlight onto Roman ships in the harbor, igniting them and saving Syracuse. Yet Archimedes failed to find an ultimate weapon to stop the siege, and the Romans sacked the city in 212 BCE, with Archimedes among the casualties.[3]

Or so the story goes. Although historical records from ancient times exist, they are incomplete, mixed with legend, and often unreliable. "Archimedes was so famous that legends clung to him," wrote Stanford historian Reviel Netz. He added that after 2,200 years, "how are we to separate history from legend?"[4]

We know that the Greeks knew that lenses or curved mirrors can focus sunlight to ignite a fire. The Greek geometer Euclid wrote about optics but mainly viewed it as the geometry of light rays. Archimedes also wrote about optics, but few of his writings survive. Most ancient Greek writings on optics are about the geometry of light rays and Greek theories of vision, which are bizarre by modern standards, considering light a form of fire that some held originated in the eyes.[5]

The only solid facts we know about the life of Archimedes are that he lived and worked in Syracuse and died there in 212 BCE when the city fell to the Romans. The standard story of his death comes from the Roman biographer Plutarch, who tells that a Roman soldier sacking the city came upon Archimedes sitting in the sand, working on a problem. The soldier demanded that Archimedes get up and go with him to meet the Roman general Marcellus. Plutarch wrote that when Archimedes refused to get up until he had solved the problem, the soldier lost his temper and killed the mathematician with his sword. Yet historian Reviel Netz says we really don't know how Archimedes died, because even Plutarch mentioned other accounts.[6]

The most famous Archimedes story tells how he jumped from his bath and ran down the street naked, shouting "eureka" because he had realized how to measure the volume of a crown by watching water spill from his bath. It's a charming story that makes Archimedes the prototype for the eccentric scientist. Yet Netz calls it unlikely because it was first mentioned as a historical anecdote in a book on architecture written some two hundred years later by Vitruvius, whom he considers "not a very reliable historian."[7]

The story of Archimedes and setting Roman ships afire is even more dubious to modern historians. Contemporary reports describe the siege and identify Archimedes as the engineer behind the city's fortifications and battle gear. Yet the earliest account of him using mirrors to burn Roman ships was written in the twelfth century by Johannes Tzetzes. "What he has to say about Archimedes comes from a gossipy, fanciful poem," Netz writes.[8] After a detailed analysis of known sources, British scholar Dennis L. Simms concluded the burning mirrors story was "assembled by a series of misunderstandings" over the centuries, based on the reputation of Archimedes and poor translations of the scanty surviving records.[9]

Yet, as the scientific adage goes, "the absence of evidence is not the evidence of absence." Just because we lack definitive evidence that Archimedes set the Roman ships afire, we can't conclude that it was impossible. We simply don't know that much about events 2,200 years ago.

We now know that the ancient Greeks had some surprisingly sophisticated machinery, such as the remarkable Antikythera mechanism recovered in 1900 from an ancient shipwreck off a Greek island. The bronze mechanism was severely corroded after two millennia in the ocean, but studies with modern instruments revealed it was a geared device that might have been a clock or a mechanical model of movement of the planets in the sky called an orrery.[10] Inscriptions on the device show it was made about 125 years after Archimedes died, and probably lost a few years later when the ship sank. It might have been a replica of a mechanical bronze sphere that Archimedes was said to have made to reproduce the motion of the planets, which the Roman general Marcellus had brought to Rome after conquering Syracuse. X-ray studies of the Antikythera device indicate it could have predicted solar and lunar eclipses as well as the motion of the sun, moon, and planets. Think of it as an ancient mechanical computer.[11] If such a sophisticated machine could be built over 2,000 years ago, isn't it worth testing if Archimedes could have ignited Roman ships?

In fact, scientists have been intrigued by the legend as far back as the eighteenth century, and some have done the properly scientific approach of experimenting. In 1747 the French naturalist George Louis Leclerc, comte de Buffon, focused an array of 168 mirrors, each about the size of a sheet of paper, on planks 150 feet (forty-six meters) away. He reported success.[12] More than two centuries later, Greek scientist Ioannis Sakkis gave about sixty Greek soldiers oblong mirrors three by five feet wide, and they ignited a wooden ship at roughly the same distance as Buffon.[13]

Proper science requires verifying experiments by repeating them, and the *MythBusters* television series couldn't resist the challenge, which they tackled in an episode that first aired on September 29, 2004. Their verdict: "In order to have any effect, the mirror would have to be impractically large, and even then, the temperature of wood only raised a few degrees. On the Discovery website, however, a challenge was thrown out to the viewers to come up with an experiment to prove it plausible, and so far, a few of the entries seem to have done so. When all the tests were completed, the myth was conclusively busted."[14]

In the fall of 2005, an MIT class in product engineering processes tried their own experiments. They set up an array of folding chairs in a courtyard, set one-foot-square mirrors in each one, and moved the mirrors so they pointed at the same spot on a wooden profile of a boat about a hundred feet away. The instructors thought the experiment might work, but most of the students were skeptical.

The sun didn't cooperate on the first day, so a few days later they set up the mirrors on the roof of a campus garage. They finally got the sun to cooperate on October 5. At first they saw some smoke but no charring. Clouds got in the way, and the sun moved, so they had to move the boat. Finally the sun reached a clear patch of sky, and they saw serious smoke. Then carbon in the wood began burning, meaning the sun-heated surface had reached at least 750 degrees Fahrenheit. Fumes began evaporating from the wood. Then, in a flash, the wood ignited (see photo insert), less than ten minutes after the sun had reached a clear patch, burning a hole through the planks before the group smothered the fire. So they announced they had busted *MythBusters*.[15]

Then MIT students and the instructors flew to San Francisco and ran a joint test on Saturday, October 22, in San Francisco, where *MythBusters* is based. They aimed their mirrors at an eighty-year-old fishing boat, which had sunk earlier but had been recovered before the test. "We were able to cause charring and smoldering in a 1–2 foot wide swath along most of the boat's length. After three passes over the boat, the hull was penetrated and a small open flame was achieved," the MIT group reported. The burning went on for a couple of hours, creating a ten-inch hole in the hull, but there was no flash ignition.[16]

Yet when *MythBusters* ran their own test in an episode that aired on January 25, 2006, they declared the "myth" re-busted: "The large scale array simply took too long to light the ship on fire. On top of that the ship only ignited when it was stationary and positioned at less than half the distance described in the myth. The myth was plausible at a smaller scale, however. Flaming arrows were fired from a ballista at the ship, but to little effect. The most effective (and plausible with Archimedes-era technology) method of lighting the ship ablaze was through the use of Molotov cocktails."[17]

What the experiments tell us is that the legend of Archimedes and the burning mirrors falls into a hazy zone. If everything is just right, it might work. The sea might have been dead calm, the sun high in a clear sky, and tar or greasy rags may have been on the deck. Put enough soldiers with shiny shields in the right place, and they might ignite a vulnerable boat. But no army could find enough soldiers with shiny mirrors to focus enough light to ignite a boat on a dark and stormy day with high seas and a low sun. So it could not have been an ultimate weapon that the Archimedes of legend could have used to burn the whole Roman fleet to save Syracuse.

But suppose Archimedes was not the eccentric scientist of legend but a wily artificer who sought merely to outwit the Romans. On a clear day with the sun high in the sky, he could have set an old boat laden with tar and tinder into the harbor, drifting out with the tide. Then, when the boat came to the right point, his soldiers could have aimed their shiny shields toward the hulk and set it aflame as it drifted toward the Roman fleet. His hope would have been that the powerful and mysterious death ray would scare the Roman commanders so they fled the harbor, hoping to escape the new weapon they did not understand or know how to defeat.

More than two millennia would pass before science would revive the death ray, and it would have the help of some wily artificers.

CONCENTRATED SUNLIGHT AND WIRELESS DEATH

The writer best known for his tales "Rip Van Winkle" and "The Legend of Sleepy Hollow" may also have invented the science fiction death ray. In 1809, before he wrote those stories, Washington Irving published a novel titled *Knickerbocker's History of New York*, in which he envisions an invasion of Earth by "lunatics" from the moon. Playing on what happened when Europeans with vastly superior weapons invaded the Americas, Irving armed the lunatics with "concentrated sunbeams." In his satirical world, those sunbeams became an ultimate weapon, overwhelming human guns much as European guns had overwhelmed Native American bows and arrows.[18]

Claims of real death rays did not come until the electrical revolution started in the late nineteenth century. In the mid-1700s, Benjamin Franklin was the first to show that lightning carries electricity, but more than a century passed before electricity was tamed well enough to transmit much power. Inevitably, creative minds started thinking about weapons. If lightning could strike with deadly force, why not electricity wielded by human hands?

Thomas Edison had not yet started work on electric lighting and the power distribution system when James C. Wingard, a New Orleans river pilot who also claimed to be a psychic medium, promised to show how a "nameless" electrical force could blow up a fifty-ton boat from afar. The self-proclaimed "professor" invited scientists, military observers, and the public to watch the spectacle on Lake Pontchartrain on May 11, 1876. He showed up late on the big day and didn't blow up anything. The committee who had helped him arrange the event wanted doctors to examine his sanity. But a month later he came back and was sitting in a boat, pointing his apparatus at the target ship a mile away when something on it exploded.[19]

The committee initially was impressed and speculated that Wingard's invention might change the nature of war. But some skeptical observers took a close look at the target ship and found a pipe with powder inside and a wire running toward the inventor's boat. It began to look as if the "professor" had faked the demonstration. Yet three years later he was still peddling his idea, with a demonstration scheduled in Boston Harbor, when a rowboat exploded in the harbor hours before the planned test, killing two men who had been trying to plant dynamite in the target ship.[20]

What brought the lethal power of electricity to the public eye was the competition between Thomas Edison and George Westinghouse to wire the United States for electricity, using different systems. Edison preferred having the current flow in the same direction all the time, called direct current, which he thought was safer. Westinghouse preferred alternating current, which reversed direction tens of times a second and could travel farther through wires. To show the dangers of alternating current, Edison staged a series of public electrocutions of dogs, calves, and eventually horses.[21] But Westinghouse won the current wars, and the prolific Edison eventually turned to other inventions.

Other inventors began working to increase voltages, most famously Nikola Tesla. Family legend held that he was born at midnight in the middle of a thunderstorm in July 1856. His family was Serbian, but he was born in Croatia at a time when both regions were part of the sprawling Austrian Empire.[22] After studying engineering in Austria and developing an interest in electricity, he moved to Paris to work for Edison's interests there in 1882, and in 1884 he sailed to America, where he worked for Edison for six months before setting off on his own.[23] Tesla had an intuitive feel for alternating current, and two of his patents became central to Westinghouse's AC system.

In 1889 Tesla began experimenting with high frequencies and their effects, which led to his invention of what we call the Tesla coil in 1890. By using high frequencies in his coils, he could build up voltages higher than anyone had yet produced, sending pulses of current through the air like artificial lightning. The high frequencies had another important effect; short pulses at high voltages could pass along the skin without injuring the body, as Tesla discovered when he accidentally touched one of his coils.[24]

Other inventors also experimented with high voltages and inevitably started wondering if artificial lightning could become the ultimate weapon. Early would-be inventors tended to be overenthusiastic. After having zapped flies with artificial lightning in 1890, a New Jersey inventor named Grinnell claimed that with a bigger and better version of his machine "thousands of soldiers can be killed in a flash, and a number of flashes are enough to destroy an army."[25] A compilation of early claims of "death ray" inventions lists several by obscure inventors and also attributes a few to Edison.[26] The reports of Edison's involvement seem unlikely. He was a prolific inventor, but at the time the press often credited him with inventions he had not made, and he usually avoided the weapons business. But those scruples did not stop Edison from paying someone else to invent the electric chair so convicted criminals could be electrocuted with alternating current from Westinghouse generators, a process Edison wanted to call being "Westinghoused."[27]

X-RAYS AND OTHER RAYS

Wilhelm Röntgen's November 1895 discovery of X-rays stimulated a wave of interest in new types of rays. Ironically, Tesla had almost discovered them first when he and a friend, Edward Hewitt, had tried to take photographs, using similar experiments. However, when Tesla first looked at the photographic plate, he thought something had spoiled it. Months later Hewitt heard of Röntgen's discovery, realized the similarity of their test, and rushed to Tesla's lab, asking to see the plate. Tesla got the plate and held it up to the light. Hewitt saw the plate showed the circle of the lens, an adjusting screw on the side, and round dots showing the metal screws on the wooden camera box. They had recorded an X-ray image of their camera. The instant Tesla saw the plate, he realized he had missed a great discovery. Frustrated and furious with himself, he smashed the plate on the floor, shattering it, muttering, "Damned fool! I never saw it."[28] He said nothing of the incident when he later described his own pioneering experiments to the *New York Times*, politely saying, "I am happy to have contributed to the development [Prof. Röntgen] has created."[29]

Today we know that X-rays are part of the electromagnetic spectrum, along with visible light, infrared light, ultraviolet light, and radio waves. In 1896 they were a mystery. Tesla told the *New York Times*, "I am getting more and more convinced that [X-rays are] a stream of material particles which strike the plate with great velocities," although at that point he thought their great speeds were "as much as a hundred kilometers a second," far short of the speed of light.[30]

Tesla can hardly be blamed for that error. Until 1799, science knew only visible light. Two years later astronomer William Herschel found his thermometer heating up beyond the red end of the solar spectrum, discovering infrared light, originally called "heat rays."[31] What we now call ultraviolet light was discovered two years later when Johann Ritter found that invisible light just beyond the violet end of the spectrum darkened silver chloride.[32] Since then the only new waves discovered had been radio waves, in 1887 by Heinrich Hertz, which were very different from light.

X-rays marked the start of a burst of new discoveries. Alpha, beta, and gamma rays followed from 1899 to 1903. All of them were energetic, and none of them were immediately understood. Physicists eagerly sought new discoveries, but a few went too far. In 1903 French physicist Prosper-René Blondlot reported a new class of rays he called N rays, after the University of Nancy where he worked. A flurry of other reports seemed to confirm the discovery, although some observers could not reproduce his work. Then American Robert W. Wood paid a visit to the Nancy lab and stealthily removed a key prism from the experimental setup, but Blondlot didn't notice the difference while making measurements, a sure sign something was amiss. Soon it became clear that the observations were flat-out wrong and N rays did not exist.[33] These days, we would suspect fraud, but a century ago observations like Blondlot's depended on the human eye, which can easily fool people into seeing what they want to see, like the famed Martian "canals" reported by astronomers of the time.

We now understand X-rays and gamma rays are electromagnetic waves, like visible and invisible light and radio waves. What makes them different from light waves is their higher energy. Alpha rays emitted by radioactive elements are particles containing two protons and two neutrons, the nuclei of helium atoms. Beta rays also are particles—electrons or positrons emitted by radioactive elements. Alpha rays, beta rays, gamma rays, and X-rays all pose hazards because they have enough energy to ionize atoms in the body, damaging tissue. Radiation poisoning can kill in the short term and cause cancer in the long term, but nobody realized that at first, and the people who discovered these new rays had not set out to make weapons.

Even the scientists, engineers, and inventors who tried to develop weapons did not claim to be making "death rays." The idea of a death ray apparently was invented by a journalist trying to explain something he didn't understand. In early 1898 John Hartman, who had served as an engineer in the American Civil War, modified a searchlight so it could guide electricity through the air. He claimed that his wireless electric gun could stun a rabbit fifty feet away and could be set to kill as well as stun. He boasted that by using more powerful light and current sources, he could guide so much electricity through the air that he could wipe out an army. That's a long way from stunning a rabbit, and his claims caught the public's imagination. William J. Fanning Jr., who wrote a fascinating book on media reports of death rays,[34] found reports of it in newspapers and magazines published in Britain, Australia, and the United States. A story headlined "The Death Ray," in the *Northern Evening Mail* published in Hartlepool, England, was the oldest use of the phrase that Fanning could find.[35] So the idea of a "death ray" was created not by a mad scientist or evil inventor but by a journalist trying to grab the reader's attention.

Hartman and his invention appear to have quietly vanished afterward. A correspondent for *The Literary Digest* could not understand how light could help electricity go through dry air.[36] Yet today's laser technology can do what Hartman could only talk about. The beam from a high-powered laser can be focused tightly enough for it to free electrons from atoms in the air, creating a path that conducts electricity. Laser beams can be used as lightning rods,[37] although at the risk of triggering a lightning strike on the laser itself, or they can be used to steer electric discharges around objects.[38]

TESLA AND WIRELESS POWER

By 1898, Nikola Tesla had become one of the superstars of the new world of electricity. He was eight years younger than Edison, but his technological charisma was a world apart from the older inventor's. Edison was a homegrown American, perpetually rumpled and unprepossessing, with a bit of middle-aged spread; Tesla was a rail-thin immigrant, with an accent, who dressed sharply. Edison was married with children; Tesla was a life-long bachelor. Edison was five feet, eight inches tall, a bit below average at the time; Tesla was six feet two, tall enough to stand out. Both tended to work intensely, but Tesla preferred the cutting edge, and Edison preferred the more practical, even venturing into milling iron ore and making concrete houses. Both were best at inventing, but Tesla was an entrepreneur and Edison more of a businessman.[39]

Figure 1.1. Inventor Nikola Tesla in his midthirties, circa 1890. (Photo courtesy of the Library of Congress, Bain Collection.)

The invention of the Tesla coil in 1890 marked Tesla's shift toward higher voltages and higher frequencies in electricity. He grew interested in wireless transmission of power and became a popular lecturer. With the Tesla coil, he could transmit enough power across a stage to light a bulb without wires. He realized that the earth itself could conduct electricity without wires, and saw the potential of sending signals wirelessly.

Tesla started developing radio-controlled automatons in 1892, but he lost his first models, years of notes and papers, and all his other equipment when a fire gutted his Manhattan laboratory on March 13, 1895. The disaster plunged him into a deep depression, which he treated by self-administered shock therapy.[40]

When he recovered, he rebuilt the lab at a new site in New York, aiming to build remote-controlled boats. Britain was building new battleships with better engines and armor in an arms race with Russia and France, with America, Germany, Japan, and Spain also developing their navies. Tesla focused on radio-controlled torpedoes that could be loaded with explosives to destroy the big new battleships. His radio controls were primitive by modern standards but were the real thing, not a fraud like Wingard's.[41]

The United States declared war on Spain while he was perfecting his boat and writing a patent, so his timing was good. So were his promotional skills. Once his patent on the radio-controlled boat was issued in 1898, Tesla promoted his new invention as an ultimate weapon. "War will cease to be possible when all the world knows to-morrow that the most feeble of nations can supply itself immediately with a weapon which will render its coast secure and its ports impregnable to the assaults of united armadas of the world," he told the *New York World*.[42]

Then Tesla turned to another passion, wireless transmission of electric power. That was a bigger challenge than transmitting telegraph signals wirelessly, which the young Guglielmo Marconi had done over distances to eighty or a hundred miles in England by late 1898. Marconi's public success bothered Tesla, who saw him as a competitor who wasn't doing anything new. After Marconi sent a message across the English Channel in March 1899, Tesla announced he had plans to send messages around the world.[43]

Tesla needed a boost. He had gone years without a big success to match the AC patents he had licensed to Westinghouse. The gossipy *Town Topics* called him "America's Own and Only Non-Inventing Inventor, the Scientist of the Delmonico Café and Waldorf-Astoria Palm Gardens."[44]

Today, it's clear Marconi and Tesla were doing two different things. Marconi powered an antenna with an electric current that changed over time, radiating radio waves that traveled through the air, spreading out and growing fainter, but still detectable more than a hundred miles away. Tesla was trying to send an electrical current long distance. That's a fundamentally different thing. Radio waves, like light, need only go one way, but they grow fainter with distance. Electric power is carried by electrons making a round trip through a complete circuit. Break an electrical circuit by turning off a switch and the electrons stop moving in the wire and the light goes off. The distinction was not that clear in 1899.

Tesla thought that he could overcome the spreading of radio waves by creating a complete wireless circuit that would send electrical current at a high frequency partly through air at low pressure and partly through the ground. The nature of that power transmission, he said in 1898, "is one of true conduction, and not to be confounded with the phenomena of induction or of electrical radiation which have heretofore been observed and experimented with."[45] That way, Tesla thought, he could transmit power for thousands of miles. In May 1899 Tesla decided to test his ideas at a new lab funded by multimillionaire John Jacob Aster IV in Colorado Springs, which offered more room for experiments at up to four million volts, less noise, and few pesky reporters asking unwanted questions.[46]

On the way across country, Tesla gave a talk in Chicago, where he discussed his wireless plans and told a reporter that he had equipment that could send signals to Mars and receive any messages that Martians sent in return. In his midforties, Tesla told the reporter his goal was "to develop a new art" that others could apply. He would be content to develop the ideas.[47]

Tesla spent the rest of the year in Colorado, performing a series of high-voltage experiments. He came back excited, claiming he had confirmed his ideas and detected messages from other worlds. Yet biographer Bernard Carlson wrote that "Tesla appears to have sought only evidence to *confirm* his hypotheses and not look for anything that might *disconfirm* his theories."[48] In other words, he cherry-picked results that would help him sell his ideas.

Talking up his results and looking for money in New York, Tesla eventually found a high-profile patron, J. Pierpont Morgan, the premier American banker of the age. The financier was interested in transatlantic wireless transmission, and Marconi had refused to sell his American patent rights. For $150,000 Tesla agreed to build a high-voltage laboratory and station for transatlantic wireless transmission. A developer donated a two-hundred-acre site on eastern Long Island with the hope of selling housing to the future staff of 2,000 to 2,500 people, and named the place Wardenclyffe. Tesla had high hopes of transmitting both signals and power across the Atlantic.[49]

Figure 1.2. Tesla built this 187-foot tower at Wardenclyffe on Long Island as a giant antenna to transmit electric signals and power across the Atlantic. The fifty-five-ton metal hemisphere was a vital part of the antenna. Most of the structure supporting it was wood. The top was visible from New Haven, Connecticut, across Long Island Sound. Tesla had wanted the tower to reach six hundred feet in the air, but he lacked the money to build that high. (Photo courtesy of the Tesla Wardenclyffe Project.)

Yet Wardenclyffe never lived up to those hopes. Tesla had promised to start transmitting across the ocean inside of eight months, but after two and a half years Wardenclyffe was not finished and Tesla had spent all of Morgan's money. Morgan had lost interest, so the giant tower stood dark as Tesla vainly sought more investors. But the problem wasn't just money. Tesla had thought that pumping electricity into the ground through pipes at Wardenclyffe would distribute power around the world because electricity flows through the ground like an incompressible fluid. However, electricity flows as a compressible fluid, so the electricity pumped into the ground would have been dissipated while passing through the planet.[50] Tesla wouldn't give up on his ideas, but the stress wore him down, and in the fall of 1905 he had another breakdown. By the time he was fifty, he felt broken by the collapse of his great hope.[51]

"DEATH RAYS" IN SCIENCE FICTION AND SCIENCE FACT

Tesla's vision of a global wireless power grid was never realized in fact, but George Griffith used it in his 1902 techno-thriller *The World Masters*, written while Tesla was working at Wardenclyffe. In the book, a global electric trust that distributes power from its base near the North Magnetic Pole in Canada threatens to wield their power to stop preparations for war in Europe. When the fighting begins, the trust uses the power beam to destroy everything made of iron and steel and disable everything electrical in the war zone. King Edward VII of Britain, outside the war zone, negotiates a truce, but in a world oddly without wireless communications, a naval fleet that does not hear of the truce steams onward in an effort to destroy the Canadian base of the network, only to be destroyed by the power beam that becomes the first fictional weapon called a "death ray."[52]

H. G. Wells armed the Martian invaders in his famed 1898 book *The War of the Worlds* with "heat rays" rather than death rays, but his description is an eerily prescient view of powerful infrared lasers, which are invisible to the eye unless the beam vaporizes dust in the air.[53]

> It is still a matter of wonder how the Martians are able to slay men so swiftly and so silently. Many think that in some way they are able to generate

> an intense heat in a chamber of practically absolute non-conductivity. This intense heat they project in a parallel beam against any object they choose by means of a polished parabolic mirror of unknown composition, much as the parabolic mirror of a light house projects a beam of light. But no one has absolutely proved these details. However it is done, it is certain that a beam of heat is the essence of the matter. Heat, and invisible, instead of visible light. Whatever is combustible flashes into flame at its touch, lead runs like water, it softens iron, cracks and melts glass, and when it falls upon water, incontinently [sic] that explodes into steam.[54]

Albert Einstein's special theory of relativity introduced the world's famous equation, $E = mc^2$, in 1905, predicting that atoms could yield tremendous energy. But decades would pass before that became an ultimate weapon in the form of the atomic bomb.

As World War I approached, a thirty-three-year-old Italian engineer named Giulio Ulivi claimed he had found a new type of invisible infrared light he called "F rays" that could detonate gunpowder wirelessly up to twenty miles away. He scored some initial successes, and some puzzling failures, during several days of tests in 1913 off the coast of Le Havre. French officials and the local US consul were impressed. He also demonstrated the trick to British officials, but nothing worked out.[55]

The next year the *New York Times* ran a long story, headlined "Invention of an Italian May Put an End to War," on Ulivi's latest exploits in what the editors seemed to hope might be an ultimate weapon. He claimed to have detonated underwater explosives several miles away so fleets and forts would be at his mercy.[56] Soon the story grew stranger. In a bizarre twist, Ulivi asked the admiral he was negotiating with for permission to marry his daughter. The admiral insisted Ulivi first show his weapon worked, but the inventor was too impatient and eloped with the daughter.[57] A few days later, another *New York Times* story reported Ulivi had faked his results by adding metallic sodium to the gunpowder in an underwater mine rigged to let water trickle in to ignite the sodium and detonate the mine.[58] Ulivi later resurfaced with claims of other inventions and further exploits of F rays.[59]

Ulivi's F rays also showed up in a novel and movie both titled *The Exploits of Elaine*, by prolific detective writer Arthur Reeve. It was one in a series of stories about Craig Kennedy, a professor who used his scientific knowledge to solve crimes, and was called an American rival to Sherlock Holmes.[60] Reeve worked real-world gyroscopes, portable seismographs, and lie detectors into his stories, and in *The Exploits of Elaine* he turned Ulivi's F rays into death rays.

In a book chapter titled "The Death Ray" Kennedy scoffs when warned "a pedestrian will drop dead outside his laboratory every hour" if he does not leave the country. When he begins investigating the next day, Kennedy takes out an instrument and says, "This thing registers some kind of wireless rays—infra-red, I think—something like those that they say that Italian scientist, Ulivi, claims he has discovered and called the 'F-rays.'" Just then, he looks outside to see a crowd gather around the body of a man who had dropped dead outside.[61]

The 1915 movie version is a classic damsel-in-distress serial, full of drama and narrow escapes, made by the director of *The Perils of Pauline* and starring the same actress, Pearl White.[62] A critic called the death-ray chapter "the best release yet" in the serial. It shows Kennedy trapped in a basement, escaping death by inches each time the death ray is aimed at him, until at the climax he uses a platinum disk to turn the death ray back on the villains and set the building on fire.[63]

The Exploits of Elaine also marks a breakthrough of sorts in fictional death rays. Instead of being physically realistic like Wells's heat rays or Griffith's power beams, Reeve's death rays killed people instantly on contact, leaving no mark on the body. That sort of death ray was more fantasy than science fiction, but it did make effective drama and was the first in a long line of handheld blasters and ray guns that eventually came to do everything from stunning to disintegrating their targets.

SEEKING WAR-STOPPING DEATH RAYS

The quest for war-stopping new weapons began soon after World War I ended. Air attacks were a major worry because they reached far beyond the battlefield. German zeppelins and winged aircraft had raided Britain 103 times, killing 1,413 people.[64] Laboratory tests showed that high voltages and intense electrical and magnetic fields could affect engines, and military leaders hoped to extend those effects high enough to affect planes. "Under the attack of these electric waves the airplane will fall as though struck by a thunderbolt, the tank will burst into flames, the dreadnought will blow up, poison gas will be dispersed," wrote French army chief of staff General Marie-Eugène Debeney in 1921.[65] In 1923 France worried that Germany had already developed such weapons after a series of French commercial flights over Germany had to make emergency landings.[66]

Military leaders hoped new science would yield new weapons, perhaps even an ultimate weapon. Technology was advancing much faster than in the nineteenth century. Relativity and quantum theory were revolutionizing physics. "We have X rays, we have heat rays, we have light rays. H. G. Wells in his *War of the Worlds* alludes to the heat rays of the Martians, and we may not be so very far from the development of some kinds of lethal ray which will shrivel up or paralyze human beings if they are unprotected," wrote British general Ernest Swinton in 1920.[67]

The press was eager to cover wild new ideas, often the wilder the better. Pulp magazines covering new science and technology proliferated, and the border between fact and fiction could be hazy. Young inventor and entrepreneur Hugo Gernsback started *Modern Electrics* in 1908, then sold that and in 1913 founded *Electrical Experimenter* in 1913, which Tesla wrote for.[68] The magazines looked toward a bright electrical future as they covered the development of radio, television, and other wonders of the age.[69] They published science fiction as well, such as George Stratton's "The Poniatowski Ray" in the January 1916 *Electrical Experimenter*, and by 1920, when the magazine's title changed to *Science and Invention*, every issue included either a speculative article or science fiction, and often both.[70] In 1926 Gernsback launched the world's first science fiction magazine, *Amazing Stories*.

"DEATH RAY" MATTHEWS

This new world eager for new technology and news about it was primed for someone like Harry Grindell Matthews when he arrived with plans for a death ray in 1924.

Born in England in 1880, Matthews served in the Boer War, where he grew intrigued by wireless communications. He saw the British army and navy experiment with the first wireless telegraphs to be used in the field, built by Marconi.[71] When he returned home to England, Matthews found a patron to sponsor his wireless research—Gilbert Sackville, the 8th Earl De La Warr, a wealthy member of the nobility whom Matthews knew from his service in the Boer War. Matthews started out to repeat earlier experiments and to extend the wireless transmission distance, but his real goal was to transmit voices wirelessly. He succeeded at that in 1907, but others had done it first.[72]

While living with his mother in 1909, Matthews filed his first patent on a portable radio telephone, called the Aerophone. With a dapper presence and an interesting idea, he raised ten thousand pounds sterling to found the Grindell Matthews Wireless Telephone Company, where he became the "superintending expert" in April 1910 at a respectable salary of five hundred pounds a year. The plan was to sell Aerophones for eighteen pounds and eighteen shillings, roughly equal to five hundred US dollars today. The technology was far short of modern mobile phones, and the company struggled. Matthews claimed the phone had a range of several miles, but he could not get a government license to transmit, and the company eventually went out of business during World War I.[73]

In December 1915, Matthews demonstrated wireless technology for remote control of a powered model boat and for exploding mines to the British Admiralty. The feat impressed officials who paid him twenty-five thousand pounds sterling, and offered another £250,000 if he could make a remote-controlled aerial torpedo. The second stage was essentially an early version of a cruise missile, and it would have been a huge advance if he had succeeded. The Admiralty investigated the technique for detonating mines.[74] The project remained secret until 1924 and gave him a military connection.[75]

After working on military projects during the war and later developing an early motion-picture sound system, Matthews had a credible track record in invention and engineering by the early 1920s. That helped him get a hearing when he started talking about energy-beam weapons in early 1924. Word reached Winston Churchill, who asked his scientific advisor to check out "the man who is said to have discovered a ray which will kill at a certain distance.... It may be all a hoax, but my experience has not been to take 'no' for an answer."[76] But once Matthews got in the door, the British military expected him to demonstrate something potentially useful.

What they wanted was not a fictional death ray that could kill people on contact but a technology that could disrupt the operation of aircraft or other military systems from afar. Today that's part of what the military calls "electronic warfare."[77] But in the 1920s, energy-beam weapons were often called death rays.

In March 1924, Matthews talked with the Air Ministry about his invention but said he was not yet ready to demonstrate it.[78] But he kept talking about it, and in May a *New York Times* correspondent had a lengthy interview with Matthews about his "diabolical rays."[79]

The inventor was elusive about the details, but he did say that the rays had been able to direct an electric current through the air for up to four miles, and he predicted that the range could be doubled. His surviving notes say the two key parts of the system were a specialized electrical generator and a carrier beam that acted as a conductor to direct the current.[80] The presence of a lens like those used in searchlights or lighthouses shows he used some form of light, perhaps invisible ultraviolet light that could ionize air so it conducted electricity. Matthews also compared the beam to a searchlight and said that he could not expect the system's range to ever exceed the visibility of a lighthouse beam up to ten miles at sea, as would be expected for ultraviolet light.[81] Electricity delivered the destructive punch. It could not destroy ships, because they were grounded by the water, Matthews said. However he claimed he could "put the ships out of action by the destruction of vital parts of the machinery and also by putting the crews temporarily out of action through shock." Airplanes, being isolated in the air, could be totally destroyed.[82] Add it up and it looks like John Hartman's plan from 1898, which might have worked with a powerful enough laser.

In May 1924, Matthews demonstrated to the British Air Ministry that his ray could stop an engine about fifty feet from the ray generator. That didn't satisfy the Air Ministry, which offered him only modest support until he could disable a small engine supplied by the government, a way for them to be sure Matthews wasn't rigging the test.[83]

The inventor rejected that proposal, then flew to Paris to discuss a possible partnership and the use of a large laboratory in Lyons where he could use a hundred kilowatts of electricity, two hundred times the five hundred watts available in his own lab. But his business partners promptly filed suit to stop him from signing any agreements.[84]

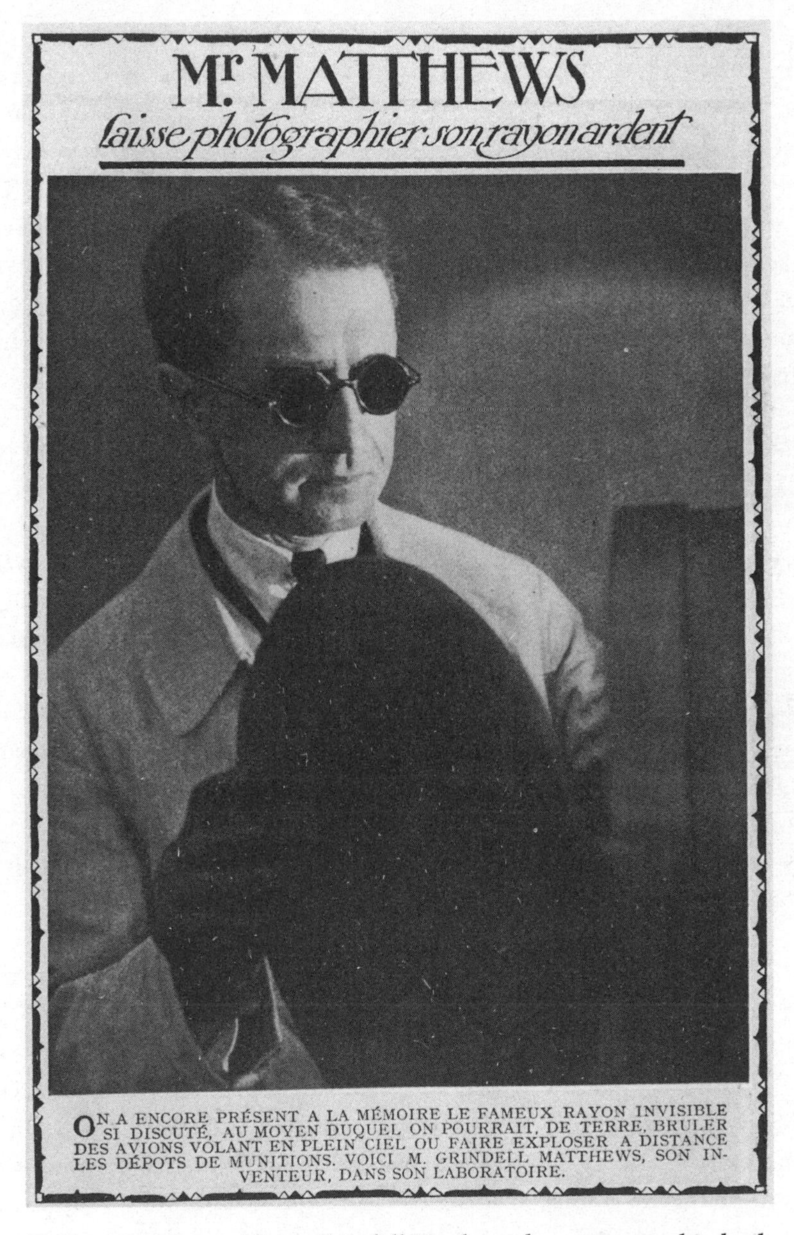

Mr MATTHEWS

laisse photographier son rayon ardent

ON A ENCORE PRÉSENT A LA MÉMOIRE LE FAMEUX RAYON INVISIBLE SI DISCUTÉ, AU MOYEN DUQUEL ON POURRAIT, DE TERRE, BRULER DES AVIONS VOLANT EN PLEIN CIEL OU FAIRE EXPLOSER A DISTANCE LES DÉPOTS DE MUNITIONS. VOICI M. GRINDELL MATTHEWS, SON INVENTEUR, DANS SON LABORATOIRE.

Figure 1.3. Inventor Harry Grindell Matthews demonstrating his death ray. (Photo courtesy of Chronicle / Alamy Stock Photo.)

In the aftermath, the *New York Times* reported that five others claimed to have invented death rays, including three in England, one in Germany, and one in Russia.[85] The paper also ran a full-page spread explaining why scientists doubted Matthews's statements. Michael Pupin, a Serbian-born physicist and engineer who headed the Department of Electro-Mechanics at Columbia University, said it was nonsense to say that any weapon could stop war. "You can't make war so frightful that men will shrink from it. . . . If a ray could be discovered which would reduce all motor-driven trucks, tanks, airplanes and battleships to scrap-iron, then armies might be obliged to go back to hand-to-hand fighting. But they would fight just the same."[86]

Pupin knew something about conflict. He and Tesla shared a Serbian heritage; they had been born there two years apart, both had immigrated to New York as young men, and both had worked in electricity and engineering. The younger Pupin had admired Tesla's early success. But the two had clashed over patents and had not spoken since 1915 when Pupin testified against Tesla's patent rights in a lawsuit.[87]

None of the criticism deterred Matthews. He kept talking with reporters and promoting his invention. Despite saying he did not like his invention to be called a "death ray," he titled a short silent film starring him and his invention *The Death Ray*. The film was shot on unstable nitrate stock, but two clips survive. An eight-minute segment shows Matthews sitting at a desk, writing, in a white coat in the laboratory, and shows the "death ray" light a lamp and explode gunpowder.[88] A fifty-five-second clip, "War's Greatest Terror," shows Matthews, neatly dressed in a suit and tie, then shows the ray detonate gunpowder and stop a fan.[89]

When Matthews visited America in October to promote his invention, he also saw a doctor, complaining of vision problems that might have come from exposure to his death ray.[90] Skeptics had begun mocking him, but for a while British authorities worried that Matthews might sell his invention to a foreign buyer.[91] Tesla, American popular technology magazines, and some reporters continued to publicize Matthews's death ray claims. But the military and most other authorities scoffed because he steadfastly refused independent testing, which might have supported valid claims.[92] After going through bankruptcy, Matthews became the fifth husband of rich Polish-born opera singer Ganna Walska in 1938, but they broke up before he died from a heart attack on September 11, 1941.[93]

TESLA IN DECLINE

Nikola Tesla remained a technological celebrity for the rest of his life, but he was never the same after his 1905 breakdown. He turned fifty the next year and never again seriously pursued any truly ambitious project.[94] Instead he lived on patent royalties, consulting fees, and income from writing.

In his glory days Tesla had lived the high life, settling in the posh Waldorf-Astoria hotel in 1898. Money became a problem later. In 1904 he gave the hotel owner a mortgage on Wardenclyffe to pay his hotel bills. The tower was torn down for its scrap value in 1917, and the property was seized to pay his hotel bills in 1921.[95] As the years passed, he lived increasingly on his reputation, the generosity of friends, and the reluctance of the Waldorf-Astoria and later the Hotel New Yorker to crack down on such a celebrity. In 1934 editor Hugo Gernsback persuaded Westinghouse to pay Tesla a monthly pension of $125 as a "consulting engineer."[96] In 1939, the Nikola Tesla Institute in Yugoslavia, worried about Tesla's health, began paying him six hundred dollars a month.[97]

As World War I raged in Europe, Tesla said he had designed wireless transmitters able to "project electrical energy in any amount to any distance and apply it for innumerable purposes, both in peace and war . . . [with those transmitters] great destructive effects can be produced at any point on the globe, determined beforehand and with great accuracy . . . the wars of the future will not be waged with explosives but with electrical means."[98]

Over the years, Tesla became the grand old man of electricity and power, sought after by newspapers and magazines for his opinions. He talked much of the future, but his ideas often echoed his past statements. An article in the October 16, 1927, issue of *Telegraph and Telephone Age* proposed a world system for wireless transmission of energy. But the concept was based on Tesla's theory that electric current can propagate through the earth faster than the speed of light in a vacuum. Tesla's mind was stuck a quarter century in the past, reinventing his dream for Wardenclyffe, not accepting that it would violate the laws of modern physics.[99]

Starting when he turned seventy-five in 1931, Tesla held large birthday parties at which he talked about the future of technology and science. Some of his ideas were out of phase with modern physics. At the first party he said he planned to disprove Albert Einstein's theory of relativity and that he did not believe that splitting atoms could release energy.[100] But he also had interesting insights. In 1934 Tesla said television "ought to be with us soon, and someday it will be on a par of perfection with broadcasting of music," and then he gestured broadly with his arm and added, "There will be large pictures thrown on the wall."[101]

At the 1934 party he also announced a new invention that he considered the most important of some seven hundred inventions he had already made. It was a new twist on the death ray, still being sought by armies. Instead of transmitting rays through the air, he envisioned sending a "death beam" of concentrated particles "as far as a telescope could see an object on the ground and as far as the curvature of the Earth would permit it." Powered by fifty million volts, higher than ever before generated, the death beams would require power plants too massive for anything but a battleship to carry, making warfare obsolete. The death beams also could detect and destroy submarines underwater, making them obsolete.[102] The beams could carry enough energy to blast a fleet of ten thousand enemy planes out of the air 250 miles from the defender's border, or kill an army of millions, leaving no trace of what had killed them.[103]

Tesla's announcement made headlines in a world in the depths of the Great Depression and fearful of war and unrest. With Edison dead, people may have hoped the great Tesla could save the world with technology. But as usual in his later years, Tesla published no more details and performed no demonstrations, leaving his ideas unclear. Only half a century later in 1984 did the Tesla Museum in Belgrade uncover a paper in which Tesla explained that his plan was to accelerate tiny particles of tungsten or mercury to high speeds because light waves could not carry lethal doses of energy.[104] It's far from clear if that would have worked.

NOBODY COULD KILL A SHEEP

Since the era of Matthews's death-ray claims, the British Air Ministry had taken a pragmatic approach to dealing with would-be inventors pitching their death-ray schemes. They offered a bounty of one thousand pounds sterling to anyone with a ray weapon that could kill a sheep a hundred yards away. Passing that simple test presumably would have earned the successful inventor the bounty as well as an invitation to talk with ministry officials, but no one did pass.[105] It might not have helped in assessing a death ray intended to stop a plane's engine rather than kill the pilot, but the ministry thought that the metal shielding could protect engines from ray weapons, so they preferred targeting the pilot.

Aerial bombing continued to worry Britain. "The bomber will always get through" whatever defense exists, warned Stanley Baldwin in 1932, between his terms as prime minister. "The only defense is in offense, which means that you will have to kill more women and children more quickly than the enemy if you want to save yourselves." Churchill in 1934 urged rearming Britain and raged against the "cursed, hellish invention and development of war from the air."[106]

In 1934, one of those small, bespectacled, and otherwise faceless bureaucrats who wind up shaping history, A. P. Rowe, determinedly dug through all fifty-three plans that had been written for air defense, and found nothing promising. He wrote a memo warning his boss, Henry Wimperts, that unless science could offer something new, Britain was likely doomed to defeat in the next war.[107]

Wimperts started exploring the options and asked Robert Watson-Watt, who operated the Radio Research Station of the National Physical Laboratory, if "damaging radiation" from a death ray offered much prospect for defense. Watson-Watt didn't think much of the idea, but he was a careful man, and to check his opinion he instructed a subordinate, Arnold Wilkins: "Please calculate the amount of radio frequency power which should be radiated to raise the temperature of eight pints of water from 98 degrees F to 105 F at a distance of five km and a height of 1 kilometer."[108] That was a circuitous way of asking how much radiation was needed to heat a pilot's blood to an unhealthy temperature.

Wilkins quickly understood the reason for the question, and his calculations showed that it wasn't practical. Watson-Watt was not surprised and asked what else radio waves might be able to do. Wilkins recalled that radio receivers often picked up noise when planes flew nearby and wondered if it might be possible to detect planes by reflecting radio waves off them. Watson-Watt jumped at the idea, and existing radio technology turned out to be up to the task. Thus the quest for death rays led to the invention of radar, which played a crucial role in protecting Britain from bombers and rockets by detecting their presence and targeting them for gunneries.[109]

Death-ray inventors didn't go away, and the press kept reporting their claims. But most reports didn't hold up or were out-and-out wrong. In 1935, *Modern Mechanix and Inventions* reported that Princeton professor Robert Van de Graaff had invented a seven-million-volt generator that could be mounted in a tank and used to fire bolts of electricity at enemies.[110] The generator did fire bolts of electricity, but Van de Graaff built it at MIT for scientific research, not as a weapon, and for many years it has made sparks fly in demonstrations at the Boston Museum of Science.[111]

DEATH RAYS AND WORLD WAR II

British military planners wrote off Matthews years before Europe started slipping into the Second World War. The inventor settled in a remote area of Wales, where he lived in relative obscurity and died in 1941. Details of his last days and the aftermath are hazy. Shortly before his death, Matthews supposedly turned down a German agent's offer of fifty thousand pounds sterling for his death-ray plans. British agents may have taken some papers from his home after his death, and American troops may have been stationed at the house later in the war, but his biographer found no official confirmation.[112]

Tesla grew reclusive and increasingly frail in his last years and died in bed at the Hotel New Yorker on January 7, 1943. One of his nephews, Sava Kosanovic, was an official in the Yugoslavian government in exile during World War II,[113] and checked his effects afterward. He found no will, but as a close blood relative, Kosanovic had a claim on Tesla's estate and removed a few photos and letters from Tesla's safe.

Then things started getting strange. Tesla's talk of death rays raised concerns that he might have sensitive documents, but the federal bureaucracy had to decide who was responsible. Tesla had long been a US citizen, but his nephew was not, so the nod went to the Office of Alien Property Custodian, which could confiscate his papers. Agents hauled two truckloads of papers from the hotel to the Manhattan Storage Company, which already had eighty barrels and boxes that Tesla had stored there years earlier.[114]

The agency then called in an MIT engineering professor who was an expert on high-voltage equipment and had worked with Van de Graaff on the high-voltage generator. After spending January 26 and 27 burrowing through all the notes and material that had been in Tesla's "immediate possession" when he died, the professor concluded there was nothing to worry about.

"It is my considered opinion that there exist among Dr. Tesla's papers and possessions no scientific notes, descriptions of hitherto unrevealed methods or devices, or actual apparatus which could be of significant value to this country or which would constitute a hazard in unfriendly hands. I can therefore see no technical or military reason why further custody of the property should be retained," wrote Professor John G. Trump, signing his letter as a technical aide to the government's wartime National Defense Research Committee, and attaching a list of Tesla projects.[115] In other words, Tesla had done little technically significant work in his later years.

Trump was well aware that his comments about Tesla might seem harsh, so he closed by saying, "It should be no discredit to this distinguished engineer and scientist whose solid contributions to the electrical art were made at the beginning of the present century to report that his thoughts and efforts during the past fifteen years were primarily of a speculative, philosophical, and somewhat promotional character—often concerned with the production and wireless transmission of power—but did not include new sound, workable principles or methods for realizing such results."[116] Those words reflect a colleague's memory of Trump as "remarkably even-tempered, with kindness and consideration to all, never threatening or arrogant in manner, even when under high stress. He was outwardly and in appearance the mildest of men, with a convincing persuasiveness, carefully marshaling all his facts."[117] So it may be surprising to find that Professor John G. Trump was the uncle of President Donald John Trump.

Trump did find that Tesla had tried to peddle the "death beams" of tiny particles he described at his 1934 birthday party. "Such beams would constitute a death ray capable of the protection of Great Britain from air attack," Tesla had written British military officials who politely declined, and Trump agreed the plan wasn't workable.[118] Tesla had received $25,000 from a Soviet trading company for plans to generate up to fifty million volts to accelerate tiny particles to high speeds. It was essentially the same proposal he had made to Britain, and the Soviets complained that they couldn't get it to work.[119]

Although death rays and beam weapons of all sorts had become standard for science fiction heroes by the early 1940s, Britain, the Soviet Union, and the United States preferred old-fashioned guns for shooting down enemy aircraft tracked by radar. Hitler banned weapon programs that could not deliver usable hardware within months, so Nazi Germany never tried to develop microwave radars.[120] Even in the last weeks of the war, top German leaders scoffed at death rays. Arms minister Albert Speer recalled labor minister Robert Ley demanding that Speer immediately launch a program to build a death ray that the arms ministry had rejected earlier. "The whole thing was so absurd that I did not bother to contradict him," Speer wrote, so he authorized the program and put Ley in charge. When the inventor sent in a list of essential supplies, Speer was amused to find that one supposedly essential part had not been made for about forty years, when it had been mentioned in a high school physics textbook.[121]

The Japanese military did try to develop powerful microwave tubes for use as death rays, and managed to kill rabbits up to thirty meters from a tube emitting two hundred to three hundred kilowatts hooked up to a giant ten-meter focusing antenna.[122] That's a few hundred times the power of a modern microwave oven, but the lethal distance was only a third of the British Air Ministry minimum for killing a sheep. Americans studying the program after the war noted that the animal had to be held in place for it to be killed, which couldn't be done with human enemies.[123]

When atomic bombs devastated Hiroshima and Nagasaki to end World War II, the scale of the killing and devastation made the atomic bomb the ultimate weapon of the nuclear age. Death rays were left to the realm of science fiction.

SCIENCE FICTION DEATH RAYS, POP CULTURE, AND OPPORTUNITY

Hugo Gernsback found a fertile opportunity when he started adding science fiction to his popular radio and electronics magazines. The fast-growing fields were attracting a young audience interested in new technology and the future of a changing world. They also were looking for entertainment, and popular fiction was booming. Pulp magazines multiplied in numbers and varieties in the early twentieth century. *Detective Story* first appeared in October 1915, and other crime and detective stories followed. Other pulps published stories about aviation, stories about war, horror stories, and in 1926 Gernsback launched *Amazing Stories*, the first magazine specializing in science fiction, and many more followed during the pulp era, which lasted through the 1940s.[124]

In part, the pulp science fiction magazines followed the tradition of the prolific novelists Jules Verne and H. G. Wells, adding invention and technology to adventure and war stories. Wells also added social commentary to his most famous science fiction novels, *The War of the Worlds* and *The Time Machine*, which also are exciting stories that continue attracting readers. Pulp science fiction magazines reprinted Wells, but they also developed their own writers with their own style, part action and part speculation about the future.

Action stories set in the future need futuristic weapons. Heat rays, death rays, energy blasters, and ray guns came early and often were depicted on the covers of magazines like *Amazing Stories* (see photo insert). The Martian heat rays of *The War of the Worlds* were something like light artillery, built into their tripod fighting suits. The simple kill-on-contact death ray of Arthur B. Reeves's *The Exploits of Elaine* tended to get more features over the years to make stories more interesting. Blasters and ray guns were handgun-style energy weapons. Tractor rays could latch onto objects and move them. Disintegrator rays did what the name implies, breaking targets into their component atoms so they seemed to evaporate or melt into a puddle. Lightsabers were swordlike energy weapons. *Star Trek* phasers could be set to kill or stun. Ray guns might fire particles, raw energy, or mysterious rays. Light cannons were heavy artillery. The list could go on.

Part of the fun for writers and readers is to envision an ultimate weapon, like the Death Star of the Star Wars movies, and to find some fatal weakness that the hero could use to destroy that weapon and topple an evil empire. Future cities or spacecraft could be equipped with force fields that in theory could block all attacks. In the real world, we would like to have a perfect force field as an ultimate defensive weapon to protect us from all threats. But in the world of fiction, a perfect force field would be boring because all the action would be futile. Action stories are no fun without vulnerabilities.

Science fiction soon moved from the pulps to the silver screen and later television. The twenty-nine-year-old Ronald Reagan starred as secret service agent Brass Bancroft in the 1940 film *Murder in the Air*, in which he foiled bad guys trying to destroy a newly developed airplane-mounted weapon called an "Inertial Projector," a defensive ray weapon designed to disable engines of motorized vehicles, much like the death rays sought in the 1920s to disable enemy aircraft.[125] Buck Rogers and Flash Gordon appeared in comic strips, comic books, and television shows. So science fiction became part of the popular culture of midcentury America. Children playing cops and robbers played with toy guns and imagined they were in the Wild West; children playing space pilots played with toy ray guns and pretended they were in the future.

We learn from playing, so toys can shape our future. A toy Buck Rogers ray gun helped set Hildreth "Hal" Walker on the path to shoot lasers at a mirror left on the moon by Apollo 11 astronauts Buzz Aldrin and Neil Armstrong.[126]

As a boy, Walker lived in rural central Louisiana in the late 1930s and early 1940s. As a young African American, his opportunities were limited, and he felt as if he were living in the horse-and-buggy era. He was curious about gadgets, the kind of kid who looked under rocks and asked, "Why?" One year he asked his father to give him a BB gun, a popular toy at the time. Instead, his father gave him the Buck Rogers ray gun, with a spinning wheel that made sparks as you pulled the trigger. It caught Walker's attention because it was more than a BB gun. "I didn't realize that he knew about this kind of thing," he recalls. "Now I look at it and say my dad really had an insight into something and he handed me the future. . . . I've always said I just followed those sparks in my life."[127]

Walker was friendly with a white boy whose family owned the local store and whose father had a side business fixing appliances. The father let Walker watch along with his son as the father took appliances apart and repaired them. Both the boys were interested, and the father would explain things, so the boys learned together and would take things apart and see how they fit together. That experience helped get Walker involved in technical things, and his interest grew over the years.[128]

His family later moved to Los Angeles, where he graduated from high school in 1951. With the Korean War in progress, he expected to be drafted. He wanted to work with technology, so he picked the navy for its training program and equipment. Four years of working with radar and electrical systems and power supplies set the stage for him to pursue engineering when he got out.[129]

That wasn't easy; discrimination was common in the 1950s, even outside the South. But eventually he landed a job at RCA, the Radio Corporation of America, where he helped build the Ballistic Missile Early Warning System, designed to warn the country in the event of a Soviet nuclear attack.[130] Then RCA put him in charge of other projects in telecommunications and radar. One big thrill was working on the first television broadcast relayed from the Telstar telecommunications satellite on July 23, 1962. Tens of millions of people watched across the United States, Canada, and Europe. Walter Cronkite opened the broadcast, many Europeans saw their first glimpse of baseball from Wrigley Field in Chicago, and viewers watched part of a press conference with President John F. Kennedy.[131] Two years later, Walker started working for one of the world's first laser companies.

TUMULTUOUS TIMES

The postwar years brought growingly tumultuous times.

Pulp magazines started declining with the paper shortages of the war years and never recovered. Costs rose, pushing publishers to shift from the ragged-edge pulp format to the smaller digest format. Pulp sales declined as television broadcasting offered new entertainment in the home. Pulps also faced increasing competition from paperbacks and comic books, and science fiction readers began shifting away from the short stories that had filled the pulps to longer novels.

The spread of television also forced changes in movies. Well into the 1940s, studios churned out large numbers of B movies, short black-and-white films produced at low cost to fill out double features in theaters. At fifty-five minutes, *Murder in the Air* was one of the shorter B movies. But that format faded away as studios shifted to longer color films and wider screens.

The legacy of the atomic bomb and the emergence of the Cold War also loomed large over the film industry. George Pal reached back more than half a century to film *The War of the Worlds*, released in 1953.[132] Orson Welles had scared America with his famous Halloween 1938 radio broadcast, but the novel had never been filmed before, perhaps because depicting the Martian invaders and the destruction they wrought was a challenge. With new film technology, Pal could show vivid images of the Martian invaders and scenes of destruction by their heat rays. He added a powerful new twist by showing humans turning to their ultimate weapon in their vain effort to stop the invaders. Bombers laden with atomic weapons swooped over the Martians, and mushroom clouds rose high in the sky after the planes passed. For a long moment, the humans thought they must have won because nothing could have survived. Then the Martians, unhurt by our ultimate weapons, strode out from the blasted area with their heat rays incinerating everything. As the Cold War intensified, it was time to find some new ultimate weapon to defend us from our deadliest creation, the nuclear bomb.

CHAPTER 2

HOW THE PENTAGON ALMOST INVENTED THE LASER

Walking into the Pentagon in early 1959, Gordon Gould was on a roll. At thirty-eight, he had come up with an idea that he knew would be the biggest thing in his life. After years of dreaming of being an inventor like Thomas Edison, Gould finally had an invention that he hoped could make him rich and famous. He called it the laser. In time the laser would make Gould a multimillionaire, perform vision-saving eye surgery, enable high-speed internet transmission, weld sheet metal, and do many other things that even Gould hadn't imagined then. But after sixty years, the laser has yet to become what the Pentagon had dreamed of then, a death ray able to zap enemy nuclear missiles into oblivion from thousands of miles away.

For Gould, it was about time for something to go right. World War II had disrupted his life, like that of many other men of his generation. He had been a Yale graduate student and instructor on his way to a doctorate in physics in 1943 when the war effort became a higher priority than preparing future professors, and he lost his draft deferral. After he passed the draft physical, the head of the physics department told Gould that he could help the war effort and stay out of uniform if he went to an address in Manhattan. On April 30, 1944, he started working for the Manhattan Project, testing a way to refine uranium so it contained the right concentration of isotopes needed to cause the runaway chain reaction that detonates an atomic bomb.[1] Thirty years later, lasers would be tested for a similar job, producing uranium fuel with the blend of isotopes needed for nuclear power plants to produce electricity.[2]

Gould loved the job, and he also found love on the job. Glen Fulwider was a brunette, pretty, vivacious, and sexy. Very sexy to a man whose two grandfathers were both Methodist ministers. Soon they were deep in a passionate affair. They also shared a passion for social justice. Gould was from a family of liberal intellectuals; his father was an editor at *Scholastic* magazine. Glen was a socialist, and when she started attending a Marxist study group led by a Polish immigrant named Joseph Prenski, in a small Sixth Avenue apartment in Greenwich Village, Gould came along. Soon both were deeply involved.[3]

In early 1945 the two protested discrimination against Jewish workers by a new company that took over their part of the Manhattan Project, which was developing a way to increase the level of the fissionable isotope uranium-235 enough to make a bomb. Within weeks, they and others were fired without explanation. They thought it was because of the protest, but the real reason was that security had heard about their Marxist leanings.[4]

Disenchanted by corporate America, Gould followed Glen in joining the Communist Party. He held a few short-lived jobs and tried to earn a living inventing. He had ideas for improving contact lenses so people could wear them longer but had no idea how to manage the business side of invention. He and Glen married but then split, and Gould drifted away from Marxism. He started teaching physics at the City College of New York and went back to grad school at Columbia. He had a new girlfriend and was pulling his life together when a New York State panel began hunting for subversives to oust from public college faculties.[5]

When they hauled Gould in, he admitted that he had been a Communist but said he had set that behind him. However, he refused to turn in the other members of the group, as the panel demanded. That cost him his job, and the world started to turn against him again. He went to tell his thesis advisor, Polykarp Kusch, that he would have to drop out. Kusch would not stand for that and said he would find university funding to keep Gould in the program. Gould moved in with his girlfriend, Ruth Francis Hill, who helped support him, and they later married.[6] Kusch, who shared the Nobel Prize in Physics, found funding for Gould. After a long effort, Gould's dissertation research finally began yielding results. But by late 1957 he was getting restless again. Then, on October 25, just three weeks after the first Soviet Sputnik launch, Charles Townes, another Columbia physics professor, called Gould in for a chat.

On the surface, the two physicists seemed similar. Both had family roots among early Puritan settlers of New England. Both were married men, with Townes (forty-two) slightly older than Gould (thirty-seven). Both were bright and could talk enthusiastically about physics and their projects. Both had dreams and ambitions.

Yet underneath they were two very different men. Townes's voice carried the tones of South Carolina where he had grown up. Gould's had the neutral tones of the New York City suburbs. Townes was proud of his Mayflower ancestry; Gould's favorite ancestor was a pirate. Townes had a home with a yard in a nice section of the Bronx and four daughters whom he supported along with his wife on his professorial salary.[7] Gould earned two hundred dollars a month as a part-time research assistant; his wife Ruth's salary as a PhD biology researcher provided most of their income, and they lived in an apartment.[8] Townes was a gentleman of the old school, proper, polite, and a pillar of the establishment. Gould was an overage graduate student who after a dozen years of kicking around felt rather put-upon by the established order. Townes had military research grants and served on military advisory panels. Gould had been sacked from a teaching job for his past membership in the Communist Party.[9] Townes socialized with the intellectual elite of New York.[10] Most important, as a full professor with an impressive discovery under his belt, Townes was the person in charge. Gould was still a student.

Townes's big discovery was that molecules could amplify radio waves without the aid of electronics. It grew from an idea that Albert Einstein had in 1916 about what would happen if atoms or molecules had extra energy. Quantum mechanics had shown that atoms and molecules could occupy a series of states with increasing energy, like steps on a ladder but not evenly spaced. Einstein said that atoms or molecules on one of the steps above the ground could release their extra energy in two ways: spontaneously without outside influence, or stimulated by a light wave having the same energy. In that second process, the light wave essentially tickled the atom or molecule so it released a second wave that was an exact duplicate of the first, with the same wave spacing and going in the same direction. Townes had shown that under the right conditions, stimulated emission from ammonia molecules could amplify identical waves spontaneously emitted by other ammonia molecules to produce a pure signal at that frequency. That's how a laser works, but the laser emits light waves, not radio waves.

The key to producing enough stimulated emission to amplify radio waves or any form of light is to put the atoms or molecules that emit the waves in the right energy state. You can again think of these energy states as steps on a ladder. Put the atoms on a higher step, and they can release energy by dropping to a lower level. Under normal conditions, fewer atoms are in higher levels than in lower ones, so atoms can drop to a lower level by themselves, but too few are on the higher steps of the energy-level ladder for stimulated emission. For stimulated emission to dominate, more atoms must be in a higher step on the energy-level ladder than in the lower step that the atoms naturally would drop to.

To understand why, we have to shift analogies. Imagine a gas made up of atoms that can be in two states, one with extra energy and one without. Suppose one atom with extra energy emits that extra energy as a photon, a quantum of light. If it encounters an atom without extra energy, the atom will absorb it. If it encounters an atom with extra energy, the photon will tickle it so it emits a second photon with the same extra energy. If most atoms have no extra energy, that photon is likely to be absorbed. But if most atoms have extra energy, the photon is likely to stimulate another atom to emit a second photon, a process that can trigger the cascade of stimulated emission in a laser.

Under normal conditions, the number of atoms in a particular state decreases as the energy of the state increases, so more are always in the lower state to soak up the light before it can trigger a cascade of stimulated emission. To make a laser, you need to put more atoms on a higher energy-level step than on a lower one, which physicists call a population inversion because it's upside down compared to the normal state. (Back in the mid-twentieth century that condition was called a "negative temperature," and physics students were taught that temperatures below absolute zero were impossible, so what Townes had done involved some serious thinking outside the box.) So people trying to make lasers need to think about population inversions and finding materials in which they can produce them.

Townes called his device the maser, for microwave amplification by the stimulated emission of radiation,[11] but he also had joked that it meant "means for the acquiring support for expensive research."[12] Now he wanted to try the same trick with light in or near the visible spectrum; he needed a way to add the right amount of energy to atoms or molecules, and he hoped a process Gould had been using in his thesis research might help. The technique was called "optical pumping" because it used light to "pump" atoms or molecules up the energy-level ladder to the desired state. It wasn't Gould's idea; as what usually happens in science, it began as someone else's idea, which other scientists adapted for their own needs. Albert Kastler invented optical pumping in France, where Columbia professor I. I. Rabi heard about it. The trick was to make a lamp that used atoms of one element, thallium in Gould's case, which emitted light at a wavelength with just enough energy to push the thallium atoms one step up the energy-level ladder. It worked like a little set of gears, with each unit of light, a photon, pushing one atom exactly one step up the energy-level ladder.[13] Optical pumping would earn its inventor, Alfred Kastler, the 1966 Nobel Prize in physics,[14] and remains a vital tool for lasers.

Figure 2.1. Charles Townes (*left*) in 1954 with an ammonia maser and his student James Gordon (*right*), who worked with him on the project. (Photo courtesy of the University Archives, Rare Book & Manuscript Library, Columbia University in the City of New York.)

Rabi knew that Gould's doctoral research had stalled, so he suggested optical pumping to get more thallium atoms into the right energy states. "Being a lowly graduate student, naturally I tried it, and that got me into optical pumping," Gould said.[15] To make it work for his experiment, he had to make a lamp in which thallium atoms would emit just the right color of light to put other thallium atoms into the right state for his experiment to work. That took Gould over a year, and success finally came late one evening when Kusch stopped in the lab to check in on his progress and lent a hand twiddling knobs on the complex apparatus. That was enough to make it all come together and work. Gould jumped and cheered, and the professor jumped with him. Then the proud Nobel laureate said, "Well, Gordon, finally," and gave his successful student an affectionate hug.[16] That made Gould Columbia's resident expert in optical pumping. He talked with Townes about using optical pumping for microwave masers, and he described his success in a lecture at Columbia. After his long struggle, Gould was finding success.[17]

Being called into Townes's office was another sign of recognition. A professor wanted his help. So Gould answered Townes's questions about optical pumping and explained how it worked. Townes, systematic and methodical, carefully recorded in his notebook that he had talked with Gould on October 25 and 28. Gould went home and thought about the implications.[18]

Townes was a microwave guy. His academic specialty was spectroscopy, studying the frequencies or wavelengths that atoms and molecules emitted and absorbed, in order to understand their interior nature. He had worked on radar at Bell Labs during World War II and had benefitted from the tons of surplus microwave equipment available after the war for research. His invention of the maser came from that research. Its purpose was precision, not power. The first maser oscillated at a pure vibrational tone of its ammonia molecules, producing twenty-four billion waves every second, with their peaks and valleys all neatly aligned as if marching in a parade.[19] It started with the first few waves that ammonia molecules emitted on their own, which tickled other ammonia atoms so they amplified that faint signal to be stronger. It might amplify the weak radar return signal so the target could be seen better, or perhaps be powerful enough to send radar pulses of its own. So he thought an optical version would do something similar, like a light bulb emitting a single pure color like a radio station transmitting on a single frequency. He wouldn't have expected a death ray.

Gould was an optics guy, who had studied optics at Yale,[20] and he had used his knowledge of optics to try to invent things.[21] Like a kid playing with lenses, he must have used one to focus sunlight to a bright spot and burnt a hole through paper. His intuitive sense of physics told him that an optical version of the maser should differ from a microwave maser like a searchlight differs from a radio broadcasting antenna, focusing light into a beam rather than broadcasting radio waves over a wide area.

To figure out what those differences were, he went back to his apartment and sat down with references and a slide rule. To a physicist, light and radio are two variations of the same thing, a wave that combines electric and magnetic fields and travels at the speed of light, which also exists as a series of chunks or quanta of energy called photons. The more energy in the wave (and in the photon), the faster it vibrates, and because the speed of light is constant, the faster the wave vibrates, the shorter the distance between adjacent peaks, the wavelength.

We sense differences in light waves as different colors. The longest red waves our eyes can see are about 75 percent longer than the shortest violet waves. The difference between radio waves and light is huge. A typical microwave is five centimeters (two inches) long, 100,000 times longer than a wave of green light. The longer the wavelength, the smaller the energy of each quantum of energy in the wave, so each green photon has 100,000 times the energy of its microwave counterpart.

Shorter wavelengths also have another advantage. Shine them through a narrow opening, and they don't spread out as fast as long waves. A rule of thumb is that light waves spread at an angle proportional to the size of the hole divided by the wavelength. So visible light would focus to a much smaller spot than radio waves.

Gould spent long hours in his study, puzzling out how to make what he called a "laser." The crucial breakthrough came when he lay in bed late one Saturday night. Put a pair of mirrors on the opposite ends of a long hollow tube full of a gas that was energized so it could amplify light like a maser amplified microwaves. Make the mirrors perfectly parallel to each other and perfectly perpendicular to the length of the tube. Have one atom emit a single photon at the right wavelength for the other atoms to amplify it. Bounce the light back and forth between the mirrors, so along the way it tickles other atoms so they emit more photons, all identical and all marching in phase, building up stronger and stronger. The waves would form a pattern so one round-trip of the cavity would be an exact whole number of wavelengths, either 100,000 or 100,001, but never 100,000.388753.

Wide awake, Gould left his wife asleep in bed, made coffee, smoked a cigarette, and sat down to make sure he wasn't fooling himself. He checked his references and made calculations and checked his references again and made more calculations. As the wee hours of the morning passed, he put the pieces together, pulled out a fountain pen, and began writing in a bound composition book: "Some rough calculations on the feasibility of a LASER."[22]

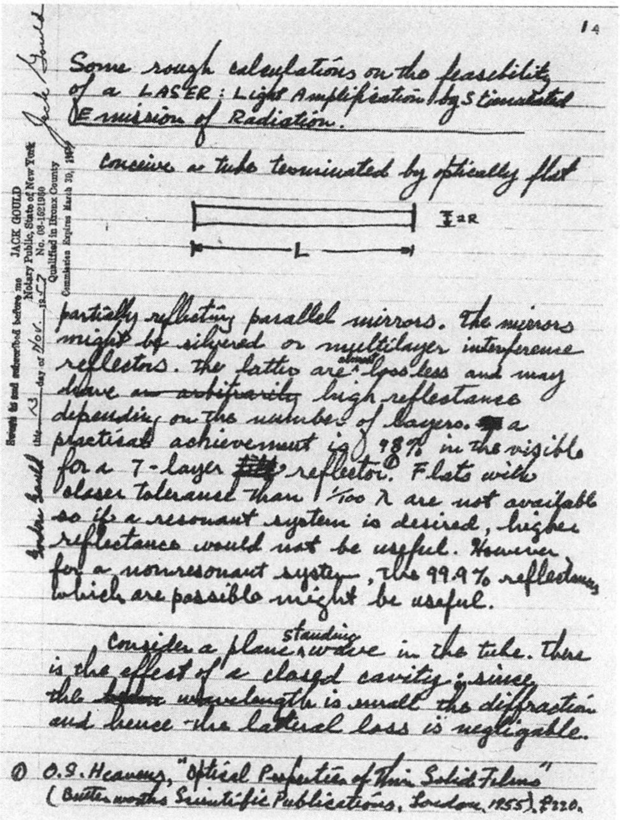

14

Some rough calculations on the feasibility of a LASER: Light Amplification by Stimulated Emission of Radiation.

Conceive a tube terminated by optically flat

partially reflecting parallel mirrors. The mirrors might be silvered or multilayer interference reflectors. The latter are almost lossless and may have an arbitrarily high reflectance depending on the number of layers. A practical achievement is 98% in the visible for a 7-layer reflector.① Flats with closer tolerance than 1/100 λ are not available so if a resonant system is desired, higher reflectance would not be useful. However for a nonresonant system, the 99.9% reflectances which are possible might be useful.

Consider a plane standing wave in the tube. There is the effect of a closed cavity; since the wavelength is small the diffraction and hence the lateral loss is negligible.

① O.S. Heavens, "Optical Properties of Thin Solid Films" (Butterworths Scientific Publications, London, 1955). P220.

Sworn to and subscribed before me this 13 day of Nov. 1957
JACK GOULD
Notary Public, State of New York
No. 03-1621950
Qualified in Bronx County
Commission Expires March 30, 19
Gordon Gould
Jack Gould

Figure 2.2. First page of Gordon Gould's notebook describing his invention of the laser. (Photo courtesy of AIP Emilio Segrè Visual Archives, Hecht Collection.)

Gould's dream since he had been in high school had been to be an independent inventor like Thomas Edison, even though he didn't know what it involved. He had realized he didn't want to spend his life at a big company, climbing endless levels of management when he worked briefly after college at the manufacturing subsidiary of AT&T when it was a telephone monopoly.[23] When Ruth woke up, he showed her his calculations and tried to explain what they meant. Hoping that he had made some progress on his long-delayed thesis, she was disappointed. She thought a doctorate was more important than a patent application, but Gould disagreed. He continued working intently until Wednesday, November 13, 1957, when he had the owner of a local candy store notarize nine pages in his notebook.[24]

He hoped it was the first step on his road to a patent, but he also knew it would be a race with Townes, who already held patents. In fact, Gould had asked Townes about the process, and Townes had explained his experience and made some recommendations.

It might have seemed logical for the two to collaborate. Both had formidable skills and expertise that would have complemented the other. Each knew the other was interested when they talked. Yet neither said anything, and each went in his own direction. Gould set out on his own, and soon afterward Townes teamed with his brother-in-law Art Schawlow to study the laser for Bell Labs.[25] In reality, a Townes-Gould team would never have worked. Townes underestimated Gould, whom he thought stole his ideas. Gould didn't trust Townes, or the establishment, and saw what he was doing as solving a physics problem that Townes had formulated.

Each also had his own agenda. Townes wanted to help Schawlow, who "had hit a flat spot" in studying superconductivity and needed a new project. Schawlow had worked with Townes at Columbia before marrying Townes's sister Aurelia, making him family as well as a respected former colleague. So Townes gave him a chance to collaborate on the laser, and Schawlow jumped at the chance.[26]

Gould was tired of grad school. He thought he should be done, but Kusch kept asking him to do more experiments. The new idea of amplifying light was far more exciting. Gould could work intensely when he put his mind to it, and unlike Townes, he already had experience with optics and light. So by November he had a clear head start on the laser. He began a second notebook in which he explored laser prospects in much more detail.[27]

Then he hit a rough spot. A discussion with a patent attorney left Gould thinking he needed to make a working model before filing, but in reality that requirement had been dropped long before except for perpetual motion machines. He dropped out of Columbia to work for the Technical Research Group (TRG), a small defense contractor, planning to spend every spare moment working on the patent. That didn't work out well, and the company president, Larry Goldmuntz, soon started asking pointed questions. But once Gould finally explained his idea to him, Goldmuntz realized that the idea could be worth a lucrative research contract. Other physicists at TRG agreed, so Goldmuntz told Gould he could use company time to write a proposal for government funding for laser development.[28]

After months of work on a 120-page proposal, they pitched two defense agencies that had funded earlier TRG projects, and Aerojet-General, which owned a minority interest in TRG. The military labs didn't think the laser would work, and Aerojet-General wasn't interested.[29] An army lab likewise couldn't be convinced that the laser would work. Such doubts were far from unusual. Goldmuntz needed a more receptive audience.

He turned to a contact who had seemed receptive to earlier ideas, although he hadn't funded any. Richard D. Holbrook invited the two to come visit him at a new organization called the Advanced Research Projects Agency (ARPA).[30] The Eisenhower administration had created ARPA in a major Pentagon reorganization after the Sputnik launch to coordinate military work on ballistic missile defense, space satellites, and other technologies the old-line agencies thought were too risky to support.[31]

The young agency that Gould and Goldmuntz walked into had a half-billion-dollar budget and was supporting top priority efforts to develop spy satellites and defense against nuclear weapons. With such a crucial role, it was based inside the Pentagon, the nerve center of the Defense Department.[32] Its charter included placing long-shot bets on crazy-sounding ideas that might just prove to be the ultimate weapon. Roy Johnson, ARPA's first director, told the House Space Committee in May 1958, "Our work might lead to a death ray. That would be the weapon of tomorrow, and obviously a man up above in a satellite would be in the position to use it." Maybe, he hinted, the death ray could replace the nuclear bomb as "the ultimate weapon."[33] Johnson didn't have any idea how to build a death ray. The former General Electric vice president was a businessman, not a technologist.[34] To solve the tough challenges posed by nuclear weapons and the space age, he told the committee, "We have to keep our mind open to everything."

Initially, ARPA's main mission was space. The US Air Force, Army, and Navy had all pitched the importance of space as the new high ground to Neil McElroy, who became defense secretary five days after the Sputnik launch. McElroy decided the best approach was to create a separate "special projects agency," and that got the support of President Dwight D. Eisenhower, a former general who had long been frustrated by the Pentagon's notorious interservice rivalry. When the dust settled, ARPA had a huge space portfolio, pieces of which would later be moved to NASA, the Central Intelligence Agency, and the armed services.[35]

Gould and Goldmuntz were not the only people thinking about lasers. Townes and Schawlow had published their less detailed laser proposal at the end of 1958 in the scholarly journal *Physical Review*.[36] Their paper would become a classic. But military contractors and research agencies didn't scour the pages of arcane research journals for new ideas.[37] Gould and Goldmuntz were the first to arrive at ARPA with a laser proposal.

Holbrook greeted them and said that he wanted someone with a broader range of expertise to listen to their pitch.[38] By profession, Paul Adams was a patent lawyer for the International Telephone and Telegraph Corporation, a major defense contractor, on assignment at ARPA because he had a deep and wide-ranging knowledge of technology and physics. But Holbrook warned that Adams also was loud and rough-edged. After walking into the conference room, he gave the two an hour to make their case.[39]

Gould did most of the talking. It was his proposal, and he had thought long and hard about what lasers could do as well as how to make them.[40] Inventions are important, but it's their applications that interest funding agencies. Gould added a natural enthusiasm that drew you into his pitch. Three decades after interviewing him in 1983 and 1984, I still recall his excitement about a then current invention, sending light down optical fibers to measure conditions at the bottom of oil wells, which is now in industrial use.[41]

Gould envisioned many uses for laser beams. With their light waves all aligned with each other, they could make accurate measurements over much longer distances than ordinary light. Lasers emitting invisible infrared light could act as powerful covert searchlights at night, revealing objects a mile away to soldiers with infrared viewers. With more power, he said, laser beams from the moon or Mars could be detected on Earth.[42]

Power had a particular allure for the Pentagon, and Gould clearly recognized the laser's potential to provide it. Lasers generate a cascade of light. It starts when an atom emits a single light wave, and that wave tickles identical atoms to emit identical light waves in the same direction. Each new light wave can generate cascades of more identical light waves, all going in the same direction. Bounce the light waves back and forth between a pair of mirrors, letting a little light escape through one mirror each time, and the resulting beam becomes powerful. All the light waves align themselves perfectly, so they all add coherently to form an intense, tightly focused beam.

Townes's radio amplifier worked similarly, but the length of radio waves made an important difference. Both radio waves and light waves can be focused down to a spot about a wavelength across, but Townes's radio waves, although shorter than most microwaves, were much longer than waves of visible light. The difference was a factor of twenty-five thousand. That meant that visible light could be focused onto an area less than two-billionths the size of one covered by radio waves. The smaller the spot, the more concentrated the power and the more intense the light. A short metal antenna on a smartphone can collect enough radio energy for you to watch a video, talk with a friend, or browse the internet. But a hand lens can concentrate visible light to an intensity a staggering 625 million times higher than those radio waves with the same power, so sunlight focused by a hand magnifier can burn a hole through paper on a sunny day.

Gould moved on to applications that needed powerful light. He showed that a laser beam could be far more intense than the brightest conventional light sources, the brilliant glow of a carbon arc used in searchlights and movie projectors. He mentioned ideas from his patent notes, using lasers to trigger chemical reactions or perhaps even trigger nuclear fusion. He went beyond the allotted hour, mentioning using light beams to guide missiles to their targets. "Maybe even knocking missiles out of the sky," he added.[43]

In 1983, looking back, Gould recalled impressing ARPA scientists and military program managers. "It came out of the clear blue that such a thing was actually possible. Ray guns and so on were part of science fiction, but actually proposing to build this thing? And he has theoretical grounds for believing it's going to work? Wow! That set them off, and those colonels, they just were too eager to believe."[44]

Adams was also sold. "I think you bastards know what you're talking about," he said.[45] But final approval rested far above him, with a panel headed by presidential science advisor James Killian.[46] Adams sent his recommendation up the chain of command. The committee's response praised Gould's proposal for its suggestions for using lasers in communications, designating targets for military missiles, and concentrating light to very high powers.[47] TRG had asked for Townes to review their proposal, and he called it valuable work that deserved funding. But he was "a little annoyed" that Gould had not mentioned the paper he and Schawlow had just published on lasers.[48]

Goldmuntz returned to Washington to get the final word, and he was happy with the prospect of a $300,000 contract; with only 12.2% of households earning more than $10,000 a year, that was serious money.[49] He was amazed when Adams added his own recommendation to the committee's report: "I, Paul Adams, think we should put $1 million into this program."[50] Goldmuntz left impressed by how well Adams understood the laser's potential impact. Very few government employees had the self-confidence to suggest increasing the size of a proposal.[51]

Adams sent the expanded proposal back up the chain of command. When final approval came in late March, the budget was an oddly precise $999,008.[52]

Everyone seemed happy. ARPA's acceptance letter praised TRG highly: "Your proposal displays keen ingenuity and it promises a vital contribution to this nation's defense effort." TRG got a contract more than triple the size it had requested. Gould got company stock options and support for his patent quest. ARPA "was exceedingly happy about the prospect of a death ray, Buck Rogers style," Gould recalled in 1984.[53]

To ARPA it was a scientifically plausible proposal for a speed-of-light energy weapon that might someday stop a nuclear missile attack. That put it within the scope of Project DEFENDER, the agency's high-priority program to defend against long-range nuclear missiles.[54] It also complemented an ARPA program called GLIPAR to explore far-out weapon concepts like antigravity, antimatter, and radiation.[55] It also could be a huge success for a new agency whose leader was embattled over issues of power and policy.[56] At the start of October 1958, ARPA had lost the civilian space program, including Wernher von Braun's group, to the newly formed National Aeronautics and Space Administration (NASA).[57] In early 1959, ARPA was fighting to retain the military space program, which would be transferred back to the armed services in September.[58]

Things were looking up for Gordon Gould at the end of March. Secretaries were typing copies of his patent application that would add up to 150 pages of details and drawings. But just days after ARPA officially approved the contract, rumors emerged that it would be classified.[59]

TRG had submitted its proposal openly, and Schawlow and Townes had already published the theory of making a laser. Groups at Bell Labs and elsewhere were trying to make lasers. So it would seem too late to try to classify the idea.

Yet it should have been no surprise. Townes and Schawlow had written about harmless lasers for communications. Gould's proposal had a keen sense of the importance of the laser's ability to generate a tightly focused beam that could reach powers to a hundred thousand watts.[60] He had told ARPA that powerful lasers might be able to destroy Soviet nuclear missiles at the speed of light, at a time when Cold War tensions were high, and generals were having nightmares that the Soviet Union was ahead in the nuclear missile race. Much of ARPA's work was highly classified, and missile defense was almost as sensitive as the top secret program to build spy satellites. Lasers fit right into that portfolio.

A worried Gould asked Goldmuntz what would happen to his patent application if the project was classified. The answer was that the patent could be filed, but it would be classified. However, classification posed another problem for TRG, which had a stake in the patents, as well as for Gould. They wanted to file patents overseas to earn more royalties. But international versions of classified patents could only be filed in three countries—Great Britain, Canada, and Australia. Gould and Goldmuntz put their heads together and talked with the lawyers, who pulled out a chunk of material they thought would not be classified, so it could be filed around the world. They left the rest in the more comprehensive version. Then the secretaries went back to typing the final versions of the two applications, which arrived at the Patent Office on Monday, April 6.[61]

What really scared Gould was dealing with the government's security apparatus. With a clearance and years of experience with the system, Goldmuntz initially shrugged off security as "a pain" because of the paperwork and bureaucracy, but not a serious threat to him. Then Gould told him about his previous problems at the Manhattan Project and City College. Goldmuntz promised to help. If the project did wind up classified, he said, "We'll get you a security clearance. They wouldn't keep you from working on your own project."[62]

Goldmuntz underestimated how deeply the security establishment had come to fear that Communist spies were hidden among American physicists, waiting to steal their secrets. After World War II many physicists who had worked on the atomic bomb had sought civilian or international control over atomic energy. At the same time, the Iron Curtain spread across Europe and Cold War tensions emerged. The 1946 conviction of British physicist Alan Nunn May for passing atomic secrets to the Soviets showed that security officers had been justified in worrying about atomic spies. The identification of more spies, and the first Soviet nuclear explosion in 1949, led to even tighter security and a growing suspicion of scientists.[63]

The rise and fall of Senator Joseph McCarthy marked the worst extremes of what became an anti-Communist hysteria. McCarthy's credibility imploded after the Senate censured him in December 1954. Yet politics and past affiliations remained an important factor in determining security clearances for years after McCarthy died in 1957. Gould was among many scientists who were sacked from faculty jobs because they refused to sign loyalty oaths or to identify others who had been involved with left-wing activities. Scientists and engineers who did have clearance were monitored and could be warned to avoid certain people or not to date certain women.[64]

TRG lawyered up after the final contract arrived, specifying that the important details were "SECRET." Security became a problem almost immediately.

Bill Culver, a Rand Corporation physicist, had invited Gould and Goldmuntz to brief the California defense think tank on possible uses of lasers. He had read the TRG proposal and thought Rand scientists needed to know about the hot new laser idea. But after they gathered in a conference room, a manager strode in, grabbed Culver's copy of the proposal, and left. Culver heard loud voices in the hall and stepped out to be berated by the company security officer because Gould hadn't submitted his clearance. Culver tried to explain that Gould had written the proposal, but the security officer would have none of it. "It doesn't matter. It's classified and you can't discuss it with him."[65]

It was the start of a sad, frustrating, and endless bureaucratic farce. Adams and TRG insisted that Gould's ingenuity was vital to the success of the project he had created. TRG hired two up-and-coming Washington attorneys to convince the Pentagon's security bureaucracy that Gould was an upstanding citizen rather than a security threat. Both would soon rise in political circles. Adam Yarmolinsky would become a Kennedy administration insider, and Lyndon Johnson would name Harold Leventhal to be an appeals court judge. But getting a clearance for Gould was far from giving a scruffy college student a shave, haircut, and suit. Gould had a checkered career as well as his political indiscretions and his refusal to turn in his former Communist friends. He didn't get along with the conservative part of the physics establishment. The highly influential Townes didn't trust him. And the security establishment was fussy about character references. The lawyers worried because a couple of Gould's potential references wore beards, and the minister who officiated at Gould's first marriage came from a "free-thinking type of church."[66]

Without a clearance, Gould could not manage the project he had created. TRG turned management over to Richard Daly, a physicist and seasoned manager whom Gould had clashed with repeatedly.[67] Gould helped recruit other people to work on it, boasting of generous funding from "Uncle Cornucopia." At a spring conference, he mentioned plans to study several different laser designs. But he couldn't talk about details.[68]

Gould's proposal had described several ways to make a laser. Once the contract was finalized, TRG asked ARPA if they should work on one approach at a time, or work on them all simultaneously. ARPA asked them to go all-out and try everything. "Even the million dollars was not enough for that," Gould said later, but TRG did set up groups to test six different approaches.[69] That meant hiring a lot of people, which took time.

Only one approach was not classified, a scheme that seemed suitable to demonstrate a laser in the laboratory but much too awkward to serve any practical or military purpose. It was a textbook example of optical pumping: shining light from potassium-vapor lamps onto another tube containing potassium vapor with mirrors on its ends to make a laser beam. The lamps were fragile, the process inefficient, and the potassium was dangerous stuff that could ignite if it hit water. But because that concept was unclassified, Gould was allowed to help a couple of young physicists working on it, as well as to work on unclassified non-laser projects.[70]

Everything else was locked down for security. Cleared TRG scientists working on those projects could ask Gould questions, but they could not tell him what they were doing or the results of their experiments, making it very difficult for him to give any useful answers.[71]

Despite the high security surrounding the laser program, the Pentagon considered it only a long shot at solving what Herbert F. York, director of defense research and engineering, called the "horrendous problem" of missile defense. During June 1959 hearings, York told a House committee that APRA planned to spend some $100 million "on research on such things as rays for intercepting missiles and on anything else that looks promising." But York called "ray" weapons "not very promising." ARPA director Johnson said the agency was investigating death rays for missile defense but admitted, "At the moment it is pretty much Buck Rogers."[72]

TRG also had nonmilitary competition in the quest for the laser. Bell Labs had started work months before Gould submitted his proposal, and had four groups trying to make lasers by mid-1959. Townes had two Columbia grad students trying to build a potassium laser based on his paper with Schawlow. Big companies including IBM, Westinghouse, and Hughes Aircraft also started looking at lasers.[73]

Those competitors began worrying about the futures of their programs as word spread that TRG's research had been classified. Bell Labs encouraged two of its groups to go public with their research early to prevent it from being made secret. Neither group had accomplished much, but both published papers on their laser plans in *Physical Review Letters*.[74] That encouraged more labs to start their own research programs.

Yet many physicists trying to make lasers got sidetracked fiddling around with unfamiliar details in optics, measurements, and materials. A TRG group charged with making a laser from transparent crystals that contained light-emitting impurities wandered off into starting a big project studying how to grow new crystals rather than starting with those already available.[75] The TRG teams working on classified projects had to spend time working out what Gould had already figured out but couldn't explain to them, because they couldn't tell him what they were doing. So they wasted time exploring dead ends.

Bell Labs hit similar pitfalls trying to make a laser counterpart of a neon lamp. The helium-neon laser was more complex because it required highly reflective mirrors on each end of a long, thin glass tube to make light bounce back and forth between them to produce a laser beam. The current had to pass through the tube in just the right way to transfer the proper amount of energy to the helium atoms so they could transfer the energy to neon atoms. The two leaders of the project, Ali Javan and Bill Bennett, were exacting perfectionists who measured and calculated everything they could before they settled down to perform an experiment. That approach made sense at the time; they didn't expect the mixture of helium and neon to generate a powerful laser beam. They thought the amount of light would increase only a little bit each time it passed through the tube, so they bought a lot of expensive equipment and carefully tried to measure those small increases in light power along the length of the tube.[76]

TRG kept hoping for Gould's clearance to come through, but nothing happened. As the company's laser expert, he should have been helping the new hires understand his brainchild. He knew the laser was going to be the biggest thing in his life, but the security bureaucracy was blocking him from working on it. He was at his best doing experiments and learning from the results, but he couldn't go in the lab and nobody could tell him the results. He virtually climbed the walls in frustration.[77]

The security restrictions became increasingly bizarre. Federal agents confiscated the notebooks Gould had filled with his laser ideas on the grounds that the contents were classified information he wasn't allowed to have, even though he had written them. The wily Gould copied them before turning them in, and it proved a wise move. After classification was lifted in 1962, government agents couldn't find one key notebook to return to Gould.[78] Even bathroom access became a problem. Working in a small unsecured area in a TRG building on Long Island, Gould had to pass through a nominally secure area to reach the men's room. At first the company looked the other way, but eventually they had a contractor tear down a wall so Gould could use the toilet without violating security.[79]

TRG continued to push to get Gould a clearance after John F. Kennedy assumed the presidency in 1961, but to no avail. Gould eventually gave up and would never receive a security clearance. The Pentagon's security bureaucracy was too powerful and too obstinate. It's unclear what the fatal problem was. In addition to his former Communist ties, Gould had rubbed some influential people the wrong way. Charles Townes was convinced Gould had stolen his ideas, and Townes had high contacts in the military establishment. In 1959, he left Columbia to become vice president and director of research for two years at the Institute for Defense Analysis, a nonprofit Washington think tank for the military, where he helped found the Jasons, a group of scientists who work together to study military projects.[80] After clashing with Gould over the TRG laser project, Richard Daly blew the whistle when he heard that Gould was having an affair with the female head of security at the company, a flagrant breach of security that got her fired.[81] Perhaps Gould's real problem was a list of indiscretions and influential enemies too long to risk giving him a clearance. In a science fiction story, some heroic figure would have found a way to break through the bureaucratic logjam and save the day. The real world is never that simple.

Whatever was behind it, classifying the laser and failing to clear Gould hobbled TRG's progress on the ARPA contract. The odds would have been against TRG even with Gould working on the laser project, but they were worse without him. Although Gould lacked Daly's management skills, he would have brought his head start on developing a laser, his understanding of the big picture, and an incentive to push ahead. The newcomers hired to work on the laser lacked his motivation and often were easily distracted by less crucial tasks they found more interesting, like growing crystals or measuring optical properties.

It's intriguing to speculate what might have happened had TRG managed to build the first working laser under the classified ARPA contract. In an alternate history science fiction drama, the United States might have used its secret laser death ray to shoot down enemy nuclear missiles and win the Cold War. But in the real world the laser could not have stayed a secret. Soviet physicist Valentin Fabrikant had proposed a laser-like device in 1939, and he and Fatima Butayeva reported some experiments in 1959.[82] Alexander Prokhorov and Nikolai Basov developed the theory behind the microwave maser in the Soviet Union in the early 1950s, independent of Townes, and later shared the Nobel Prize in physics with him.[83] Schawlow and Townes outlined laser theory in their late 1958 paper. Others used that information to make lasers without benefit of security clearances or ARPA funding. But it has still taken us six decades to reach a point where our laser firepower is approaching the level needed to shoot down rockets far less threatening than nuclear missiles.

Gould's personal story had even stranger twists. He secured patents in Britain, Canada, and Australia in the 1960s, but US courts ruled against him. Hoping against hope, he kept fighting, and finally he was able to secure a series of four patents on lasers in the late 1970s and early 1980s. Later court cases sustained his case.[84] Because the laser industry had grown tremendously by then, I estimate Gould earned patent royalties in the tens of millions of dollars, far more than he would have earned had his initial US patent succeeded.

Figure 2.3. A happy Gordon Gould shortly after his four laser patents were issued. (Photo courtesy of AIP Emilio Segrè Visual Archives, Hecht Collection.)

A RACE TO MAKE THE LASER

Publication of the Schawlow-Townes paper had lured others into what became a race to make the laser. Classification kept TRG quiet on its progress, but it had little to report. Bell Labs was thought to be in the lead, with multiple groups working on laser projects, and had reported some progress. But in reality progress was slow. Ali Javan and Bill Bennett had bought a lot of expensive instruments in their effort to make a laser from a mixture of helium and neon, but Bell Labs management was worrying that no laser light would ever emerge from the money pit.[85] Schawlow once had hope for laser emission from synthetic ruby, a crystal containing chromium that emitted deep red light when bright yellow light shined on it. However, he dropped it because other researchers had reported that ruby emitted much less light than it absorbed.

One of several outsiders interested in making lasers, Theodore "Ted" Maiman at Hughes Research Labs, wondered what had gone wrong with ruby. He had used ruby crystals to make a big advance in microwave masers, and it had worked so well that Hughes had given him time to look for another idea that might land another military contract. He needed to figure out where the lost light went so he could find a better laser material.

Maiman started out to be an electronic engineer, like his father, but after serving in the navy and getting an engineering degree from college, he shifted to physics and earned a doctorate from Stanford under Willis Lamb, who had worked at Columbia and shared the 1955 Nobel Prize in physics with Gould's advisor, Polykarp Kusch. Maiman's engineering skills nicely complemented his physics training in the laboratory. The laser idea interested him because he wanted to explore physics rather than just build another device.[86]

The key to winning the laser race was to produce lots of atoms that would sit around with just the right amount of extra energy waiting to release it in a laser beam. That required finding the right material, and Maiman had thought ruby should be good from his earlier experiments. However, other scientists had said the crystal was not releasing enough light, so he needed to measure where the yellow light focused into the ruby was getting lost so it couldn't produce red light. To do that, he needed an instrument called a monochromator, which could generate just the right color of yellow light to measure the crucial efficiency. That cost

$1,500, enough to buy a good used car at the time, and had to be approved by the department manager, Harold Lyons, who agreed only reluctantly.

The new instrument soon proved its worth. Maiman thought the crystal was somehow scattering the incoming yellow light. But when he and his assistant carefully set up the monochromator and a bunch of sensors, they saw no sign of yellow light being scattered away from the ruby. In fact, the ruby was soaking up nearly all of the incoming yellow light, then emitting red light. Ruby might make a good laser after all.[87]

Lyons was far from enthusiastic about the idea, but Maiman decided ruby crystals were worth trying and charged off on his own. Maiman could be a bit headstrong at times, and he didn't get along well with Lyons. He was not the only one. It was a common problem in industrial research labs, and Hughes was a peculiar type of lab. It was a subsidiary of Hughes Aircraft, a military contractor whose sole owner was the rich and increasingly eccentric and withdrawn Howard Hughes.

The research lab had started in an old aircraft factory in Culver City, California, but Hughes was preparing to move it to a shiny new hillside building in Malibu with a stunning view of the Pacific. Scientists ran the place, and even the managers never saw the reclusive Hughes. It was a busy and creative place, specializing in cutting-edge research and funded by military research money. Hughes managers invited eminent speakers to stimulate their scientists. Richard Feynman from Caltech was a frequent speaker as well as a consultant for Hughes Research Labs.

Maiman rarely attended Feynman's talks. He avoided Lyons; it was the easiest way to deal with a difficult manager. He focused laser-like on his plans for making a laser. He knew that would require a very intense light to excite most chromium atoms in the ruby, and when he sat down to calculate how bright, he found the intensity he needed was just within reach. With his background in electronics, Maiman understood a crucial factor in laser operation was the feedback provided by light reflection from the laser mirrors. The more light is reflected back through the laser, the faster the laser power grows. It's like the sudden runaway screech you hear in an auditorium when a misdirected microphone picks up the sound from a speaker. That sort of feedback is bad news in audio systems, but it's just what you want to increase the power in a laser beam.[88]

To look for suitable lamps, he turned to a part-time assistant, Charles Asawa, a graduate student at UCLA who was almost forty, six years older than Maiman. He was born in Southern California in 1920, where his Japanese immigrant parents had a truck farm. He was interested in physics and math from an early age, but after the deaths of his father and older brother he was needed to help his mother with the farm. He was taking night classes at a community college when Japan bombed Pearl Harbor, and the family was interned with other Japanese Americans. He was later drafted and served in the army as a translator in occupied Japan. Asawa finished college and worked for several years before he started graduate school.[89]

Movie projector bulbs came close to what they needed, but they were expensive and short-lived. After looking carefully at the numbers, Maiman decided that even the brightest projector bulbs would be marginal, and he wanted better performance than that to convince other physicists that he really had a laser. That meant turning to a lamp that could be pulsed to emit much brighter light in a brief burst, so he asked Asawa to look for bright pulsed lamps. Asawa mentioned the quest to his office mates, and one who was an avid amateur photographer showed Asawa his latest expensive toy—a coiled photographic flashlamp. That was an innovation at the time; most home cameras used cheap flashbulbs that fired only once and burnt out. The coiled flashlamp was brighter and could be fired many times. The spring-shape coiled lamp also had a special advantage for their laser experiments. They could stick a ruby rod inside the coil, so when the lamp was flashed, its light would pour into the ruby rod from all directions. Maiman ordered several of each of three models of the coiled lamps and began planning an experiment.[90]

Hughes Research Labs had been moving one group at a time from the old plant to the new laboratory, and Maiman's time came as he was planning the experiment. Stuck at home during the move, he wrote a paper reporting that he had seen bright emission from the ruby and submitted it to *Physical Review Letters*.[91] Unpacking and reassembling scientific equipment is a slow and careful task, so the moving process took a few weeks. Maiman now had a long, narrow office with an ocean view, but his mind was elsewhere, in a windowless laboratory across the hall, where he and his full-time assistant Irnee D'Haenens were setting up the laser experiment. They found a power supply that could generate short pulses of voltage adjustable to more than one thousand volts to fire into the flashlamp. The plan was to turn up the voltage in a series of steps. The higher the voltage, the brighter the flash of light, so eventually, if all went well, one flash would be bright enough to push the laser past the threshold and it would fire a burst of red light.

Maiman started with the smallest of the three coiled flashlamps. He calculated that it should flash brightly enough to cross the laser threshold. If it worked, the red light from the ruby would spike, much brighter and much sharper in time than the light from the lamp. If it didn't, it would be on to the midsize lamp.

They found a little ruby cylinder, about the size of the tip of a little finger, and had its ends polished smooth and coated with shiny silver to reflect light back into the cylinder. Then Maiman scraped off part of the silver on one end, so the beam of laser light could escape. They slipped the stubby rod inside the coil of the lamp, then slid the two into a reflective aluminum cylinder that the Hughes machine shop had made to hold the assembly. The housing would reflect light from the outside of the coil back toward the ruby inside the lamp, and would contain most of the light from the lamp so it didn't flash blind the two. The metal housing with the lamp and laser rod inside was smaller than an adult's open hand.

To measure the light, they used a standard set of electronic tools. They pointed the output from the metal cylinder at a light-sensitive vacuum tube, which generated an electronic signal proportional to how much light it saw. The tube was very sensitive, so they blocked part of the light they hoped to see from the laser. They also put a filter in front of the tube to keep the white light from the lamp from overwhelming the red light from the ruby crystal. The sensor couldn't tell the difference between the colors, and a little light from the lamp inevitably got through, but they hoped the filter would block enough white light to measure the red ruby light.

A wire carried the electronic signal from the sensor to the all-purpose tool of every electronics lab, an oscilloscope. It was a big metal box, about the size of an old-fashioned black-and-white tube television, with metal on the front surrounding a small picture tube several inches across with a square grid of calibration lines marked on the front. Circuits inside the box controlled a beam of electrons inside the picture tube so they swept curving lines across the screen, displaying the rise and fall of an electrical signal over time. Electric pulses both fired the lamp and triggered the oscilloscope to scan electrons across the screen and display the shape of the measured pulse. If it looked interesting, they would take a snapshot with a Polaroid instant camera built specially for recording oscilloscope traces, another piece of standard lab equipment.

They started the experiment on Monday, May 16, 1960, with pulses of five hundred volts. The plan was to see what happened as they turned up the voltage. Maiman stared at the oscilloscope screen, watching to see how the shape of the pulse changed. At low power, he expected only long, low humps of lamplight that had passed through the filter. At somewhat higher power, some red light from the ruby should start to appear during the pulse. Cranking the power high enough to make a laser pulse should produce a very short sharp spike of red light.

The first pulses showed the expected low hump. Maiman anxiously watched the waves showing the results on the oscilloscope screen. His earlier experiments had made ruby look good. His calculations had predicted it would work. But uncertainty was inherent in experiments. The ruby crystal might not be good enough to make a tightly focused beam. Perhaps he had missed something that would prevent laser oscillation. He may have worried about another clash with Lyons, who hadn't wanted him working on ruby.

Each step up the voltage scale made the lamp flash a little brighter and the ruby crystal emit more red light. At that level it was technically fluorescence, like the visible colors you see when ultraviolet "black light" shines on certain minerals. A little fluorescence peak rose above the broad hump of white light. Stray light was dazzling their eyes, leaving both men partly flash blinded. Maiman kept his eyes on the oscilloscope screen, where he hoped to see a sudden sharp peak of laser light.

The change came at 950 volts. A brief spike of red laser light rose high above the hump of lamplight. Its sharpness showed the burst of laser light was much shorter than the pulse from the lamp, just what Maiman had expected. D'Haenens saw the red laser light hitting a cardboard screen and whooped and jumped with exuberance. He was color blind, and the laser light was so bright that the few red sensors in his retina could see red for the first time. So the experiment's success had a second, special excitement for him.

Maiman, older, more reserved, and more intense, was numb and emotionally drained. He was relieved to have won a high-stakes race. He knew others were after the same prize, although no one outside of Hughes yet realized how much progress he had made. He knew he would have to do more experiments and extensively document his work to convince the physicists who had said that ruby wouldn't work.

Word spread quickly in Lyons's group, but Maiman was cautious. He told Lyons that he needed another instrument to verify that he had made a ruby laser before he'd announce anything. Delighted to have such a breakthrough in his group, Lyons pulled rank to requisition the instrument from another Hughes scientist's lab. Only after he and Asawa made the key measurements of the ruby spectrum did Maiman let word spread around Hughes.[92]

What Maiman had achieved was a breakthrough made possible by his engineering approach. Everyone else was trying to make lasers that emitted a continuous beam, but Maiman couldn't find a good enough steady light source to power his ruby laser; he turned to a pulsed flashlamp to generate the high power needed for the laser. He could readily buy the lamps and most of the other equipment required. Today, engineers call that using "commercial off the shelf" technology and recognize it as key to achieving goals quickly and inexpensively. Maiman pioneered it and succeeded in making a laser for a small fraction of what ARPA paid TRG or what Bell Labs spent on the neon laser.

The next few weeks were intense. Hughes Research Labs was keenly aware of the competition to make a laser and worried particularly about Bell Labs, so they wanted to report their success quickly. That meant no time to wait for a better ruby crystal that would project a brighter, tighter spot of light on the wall than the C-shaped blur from the original laser. Maiman wrote up his results and ran it through the company chain of command, finally airmailing it to the country's premier journal for quick publication of hot physics research, *Physical Review Letters*. They had just published his earlier paper reporting a modest advance; surely they would publish his breakthrough paper on the laser.

Maiman was astounded a few days later to receive a summary rejection from the journal's editor, Samuel Goudsmit, who was bored with microwave masers, annoyed with submissions of a series of results, and had clearly missed the point. He also would not listen to Maiman's protests. With that option closed, the best option to establish priority quickly was a short letter in the British weekly *Nature*. Maiman dashed off three hundred words that gave few details, airmailed it, and got a quick approval.[93]

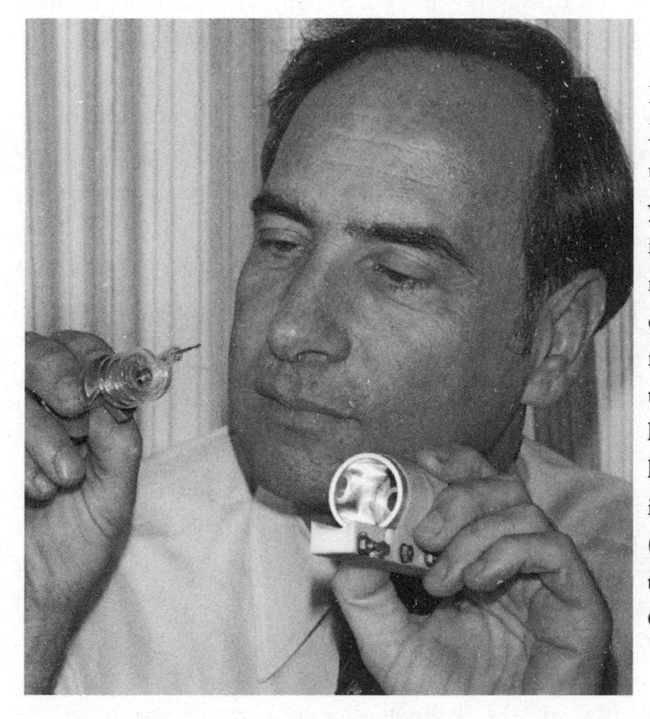

Figure 2.4. Theodore Maiman displaying the first laser, many years after his invention. The little ruby rod is inside the coiled flashlamp in his right hand; it fit into the metal cylinder in his left hand. Part of his genius was making it easy to make. (Photo courtesy of the Union Carbide Corporation.)

Lyons had begun planning a press conference after Maiman had told him about his success. It became a top priority when another Hughes manager returned home, saying that Townes's students at Columbia were close to having a laser. He would turn out to be wrong, but Hughes didn't want to take any chances and scheduled a press release in Manhattan on July 7. Then Lyons decided he should be the one to fly east to break the news.

Maiman was furious with Lyons. Intense and ambitious, he was proud of his achievement, and he wanted to be at center stage to make the announcement. The long-simmering tensions escalated quickly to a major confrontation. Word of the clash quickly rose up the management ladder. The lab's three top managers stood solidly behind Maiman. He went to New York. Lyons left Hughes shortly afterward.[94]

In his prepared talk, Maiman said that lasers might someday be used in scientific research, as well as in medicine, chemistry, machining, and transmitting signals through space. He did not mention military applications for lasers, but death rays quickly came up when reporters walked up to him after he stepped down from the podium. A reporter from the *Chicago Tribune* said, "We hear that this laser is going to be a weapon."[95] Maiman told him that wasn't in the press kit, but the reporter kept pressing his question. Trying to dissuade him, Maiman said that any such thing would be at least twenty years away. The reporter pressed harder, finally asking if Maiman would say the laser could never be used as a weapon.

Maiman's laser was far from a death ray. It was so small he could hold it in one hand. Its red pulses might have injured a gnat unfortunate enough to be sitting in the path of the beam. Yet Maiman also knew that laser technology would improve in coming decades and felt a scientist had the duty to be honest. "I can't say that, but . . .," he admitted, and was surprised when the reporter said that was all he needed and dashed off.[96]

The next morning the laser was a headline story in newspapers around the world. The story on the front page of the *New York Times* was a sober account of the promising new discovery. But what stuck in Maiman's mind forty years later was a two-inch-high red headline on the front page of the *Los Angeles Herald* and others that "were some variation of 'LA Man Discovers Science Fiction Death Ray.'"[97]

Maiman's success stunned other laser developers. Outside of Hughes, he had kept his laser work to himself until the press conference. He had reported how efficiently a ruby could emit light, but even Schawlow, who had seen that report on ruby and recommended its publication, hadn't realized that it meant a ruby laser was possible. Some doubted that Maiman had actually produced laser action.

TRG was the first to confirm that the ruby laser worked. They didn't have full details on Maiman's laser but had been considering ruby and had samples in their lab. Ron Martin in the classified program put together a laser based on a press release photo, which showed a lamp and laser rod larger than those Maiman used because to the Hughes photographer they looked better. A key to their success was cooling the laser with liquid nitrogen, which Hughes did not do. Daly called Schawlow to report their success, and within a few days Bell also had a laser running.[98] Others followed.

At first it wasn't easy. "It was a brute-force thing; you had to have a very powerful xenon flashlamp to make it work, because the ruby crystals weren't high enough quality," Martin recalled.[99] Maiman had to wait for a second ruby crystal before he could produce a pencil-thin beam.[100] Initially the only commercial ruby crystals were made for watches, but the makers improved the optical quality after Maiman's ruby laser. "In a year or two any high school kid could make a ruby laser" if he or she had a powerful flashlamp and power supply, Martin said.[101]

The laser age was off and running. Hughes quickly landed a contract to explore using lasers for measuring the range to military targets on the battlefield.[102] TRG landed another ARPA contract to continue developing lasers. By the end of 1960, IBM scientists had demonstrated two new types of lasers similar to Maiman's but using other crystals, and Bell Labs had gotten its neon laser working. Military agencies were getting excited by the potential of the incredible laser. Science fiction writers started to arm their heroes with lasers instead of death rays.

Early lasers were far from death rays. Bell Labs wanted laser communications, so two engineers hauled one of the first ruby lasers to the top of a microwave tower one night and pointed it at the roof of another lab twenty-five miles away, where a third engineer could see red flashes on a movie screen.[103] Research scientists wanted lasers to study the nature of atoms and molecules. Engineers fired laser pulses at things to see what happened, and they soon measured laser pulse energy in "Gillettes," the number of razor blades the pulse could penetrate. But those lasers could not have killed anything much larger than a fly.

CHAPTER 3

THE INCREDIBLE ROCKET-ENGINE LASER

The laser was born at a time of wild optimism about new technology, tempered by fear of nuclear doom. On May 25, 1961, President John F. Kennedy announced plans to land an astronaut on the moon by the end of the decade.[1] Electronics and computers were advancing rapidly. Nuclear power promised cheap energy.

The popular press saw the laser as part of a bright new future. Just two weeks after the end of the Cuban missile crisis, the Sunday newspaper supplement *This Week* carried a feature titled "The Incredible Laser: Death Ray or Hope" that reported,

> The laser may have greater impact than any discovery so far in the burgeoning field of electronics, which has already brought us radar, transistors, satellite tracking networks, TV. The technological revolution it brings about may dwarf any in the past.[2]

The feature, with illustrations of laser cannons straight out of science fiction, was a bit much for Arthur Schawlow, who had moved west to teach at Stanford. An inveterate punster, he pasted a copy on his laboratory door, with "For Credible Lasers, See Inside" written beside it.[3]

THE LASER BOOM

The simplicity of Maiman's design made the ruby laser easy and inexpensive to build. When the head of the Air Force Communications Sciences division expressed interest in the laser, First Lieutenant C. Martin Stickley and his boss, Rudolph Bradbury, were enthusiastic. With an excellent ruby rod made by another team at the Air Force Cambridge Research Laboratory along Boston's Route 128,[4] they started working on the laser in September and had it operating by November. "I recall requesting $392 for the purchases of capacitors and flash lamps. My request was immediately approved as everyone was excited about the prospects of having an operating red laser!" Stickley wrote later.[5] His mission was not to build death rays but "to understand why lasers did what they did," and what could be done with them.[6]

The bright red flashes from the first ruby lasers were far too feeble to be weapons, but they offered ways to improve the accuracy of conventional weapons. The accuracy of artillery depends on knowing the distance of targets, and that could be found with a laser range finder that fires a pulse and measures how long light reflected from a target takes to return to the laser. Soon after Hughes announced Maiman's laser, the air force gave them a contract to build one.[7] The Army Office of Ordnance Research had a different idea, using lasers to project bright spots to designate targets for guided missiles, and after they heard about the TRG laser program, they asked ARPA to declassify it so they could use the laser in their program.[8]

Meanwhile, ARPA and the armed services set their sights on higher-power lasers for weapons.

BIGGER AND BETTER LASERS

One approach was making bigger ruby lasers. TRG was working on it. Hughes called its project a "kill-a-rat" laser.[9] Another approach was to find other solid-state laser materials that might be larger or more efficient than ruby.

Gould also had proposed other ideas, including powering lasers by running electric discharges through a gas-filled tube, like in fluorescent lights, but with mirrors on the ends to make a laser. Bell Labs made a laser from a mixture of neon and helium that way, but it only emitted a few feeble milliwatts.[10]

Crystals like ruby had an important drawback; their size was limited by the need to grow them by a slow and painstaking process to keep the atoms properly arranged. Bell Labs came up with a new class of crystalline lasers in which the light emitter was the rare earth element neodymium.[11] But crystal growth remained an issue. Then a physicist working for a century-old manufacturer of spectacles and microscopes based in south central Massachusetts made a laser with light-emitting neodymium embedded in glass.

The American Optical Company was an old-line optics company that had created a new laboratory to develop new products. They had hired Elias Snitzer to help find new uses for optical fibers, which at the time were mostly used in medicine. Snitzer had also gotten interested in lasers, and in 1961 he made a laser using light-emitting atoms in thin glass rods that acted as optical fibers.[12] That opened a new doorway to high-energy lasers because glass could be cast into pieces much larger than the crystals used in lasers.

Like Gould, Snitzer had his career derailed by anti-Communist investigators. Born in Lynn, Massachusetts, in 1925, Snitzer became deeply involved in left-wing politics and protests as a grad student at the University of Chicago. After finishing his doctorate in 1953, he worked in industry before landing a teaching job at the Lowell Technological Institute in 1956. Then the House Un-American Activities Committee subpoenaed him for 1958 hearings in Boston. "I refused to testify on the grounds of the First Amendment concerning my political beliefs and associations. There was a little bit of bombast in what I did," Snitzer admitted in 1984, but he said he would have done the same thing if he had to do it over again. Lowell fired him for that.[13]

He spent a while kicking around. He consulted for companies including High Voltage Engineering, the company John Trump and Robert van de Graaff founded after World War II. Snitzer landed a job at MIT, but was released when the security officer returned from vacation and said he had to go. He almost landed a job at Harvard but again ran into political opposition.[14]

American Optical was a different case, a staid, old company that had joined the postwar technology boom. It had developed wide-screen movie optics for Mike Todd, who used it to produce the motion pictures *Oklahoma!* and *Around the World in 80 Days* before dying in a plane crash.[15] The company had also launched one of the world's first fiber-optic technology programs, and laboratory director Steve MacNeille was looking to replace the head of that program who had left to start his own company.[16] American Optical had some military contracts that required a clearance, but MacNeille wanted to hire Snitzer anyway, and the Pentagon okayed that as long as he stayed away from classified material. It was an awkward situation, but with his wife pregnant with their fifth child, Snitzer was happy to have a stable job.[17]

The company gave Snitzer the freedom to explore new fields, and one of them was the laser, which he thought might work well with fiber optics. The simplest type of optical fiber contains two layers of glass, a central core made of one type of glass surrounded by a concentric cladding made of a different type of glass. The trick is to select the glasses so that light traveling though the length of the core will bounce back into the core if it hits the edge of the cladding. That confines the light to the small core, preventing it from passing into the cladding and escaping from the fiber. Snitzer figured that if he put a light-emitting element in the core of the fiber, the light it generated would be trapped in the core and would bounce back and forth between mirrors put on the ends to make a laser. Snitzer tested this by winding the fibers into tight coils and putting them near lamps to see if they would reach the threshold for lasing.[18] Later he switched to fibers thick enough to be stiff rods, but which had cores like normal fibers.

Snitzer started looking for something that emitted visible light to put in the fiber core. He turned to flashlamps after he heard of Maiman's ruby laser. When he couldn't find a good emitter of visible light, he turned to elements that emit infrared light. The elements he started with emitted only weak light, but when he got to neodymium, it went off the chart with bright light.[19]

The first time he tried to make a laser, with a one-millimeter core inside a three-millimeter fiber rod, Snitzer produced so much power that it blew the end off. Soon afterward, he was talking about glass lasers with the Office of Naval Research and visiting the Institute for Defense Analyses. He told them they didn't need fibers to make glass lasers; if they wanted to get a lot of power out of it, they could make bigger rods.[20]

That was good news for laser weapon developers. The size and optical quality of crystal rods limited their ability to generate laser power. But glass rods could be made much larger, promising the much higher performance needed for weapons. [21] That was an important step.

A TIPPING POINT FOR LASER DEVELOPMENT

In December 1961, Bill Culver, who had moved to Washington to become a resident laser expert for the Institute for Defense Analyses, summed up the state of the art in lasers: "Current [laser] developments have led a number of people in government and industry to believe it may be possible to generate and direct enough coherent optical power to make a useful radiation weapon." A number of groups were working on laser weapons, and they were looking good compared to two other candidates for "radiation weapons," high-powered microwaves and beams of charged particles. Beams of laser light wouldn't bend in the earth's magnetic field like charged particles, and they wouldn't require the massive antenna needed for microwaves.[22]

The institute had a panel of experts, including Townes and Culver, spend four days between Christmas and New Year's analyzing the prospects for laser weapons. "Most people came with the idea it was crazy. They wanted to stop it," recalled Culver. However, the government's two big nuclear weapon laboratories, Los Alamos and Livermore, pointed to a potentially fatal flaw in the outer shells that kept nuclear weapons from burning up like meteors as they entered the atmosphere. They were sensitive to thermal shock, so the sudden heating from a short, powerful laser pulse could shatter them like an icy glass dropped in boiling water. That would expose the nuclear weapon to the intense heat of reentry, and it would break up in the atmosphere without exploding.[23]

Nuclear-armed intercontinental ballistic missiles (ICBMs) were the biggest and worst threats of the Cold War, able to reach their targets in half an hour. Nuclear bombers flew at subsonic speeds, so they took many hours and were vulnerable to fighter jets and antiaircraft missiles. That was why the Pentagon worried so much about Soviet nuclear missiles. If lasers could shatter the protective shells of nuclear warheads, they would become the ultimate defensive weapon, trumping nuclear missiles.[24]

At lunch, the conversation turned to the debate over the hydrogen bomb, the ultimate weapon of the time. J. Robert Oppenheimer, then head of the Atomic Energy Commission, had forcefully argued against building the "Super," but Edward Teller had won the debate and gotten the go-ahead. Afterward, Teller and other critics attacked him, his clearance was revoked, and Oppenheimer was thrown out of the nuclear establishment that he had helped create. "Nobody wanted to get into that kind of trouble," Culver recalled. So later, when Eugene Fubini, deputy director of defense research and engineering, went around the table, grilling the experts on their thoughts, everyone agreed that they should look into it. "We all were surprised how unanimous it came out. Nobody wanted to stick their neck out and be the lone dissenter," Culver said.[25]

The panel also recommended focusing on lasers based on solids because the closely packed atoms can both collect and emit more energy in a tightly directed beam than the widely dispersed atoms in gases. Up to that time, gas lasers had been limited to very low powers.[26] They pointed to Snitzer's glass laser as particularly promising for scaling to the high power needed for missile defense. They thought the chance of success in a decade was "reasonable," but nonetheless much less than half.[27]

On January 2, 1962, some of the panel presented their recommendations to Harold Brown, deputy secretary of defense for research and engineering, in his Pentagon office. Brown approved them, and allocated $5 million to ARPA to create Project SEASIDE to develop bigger glass and crystalline lasers. Contracts went to Hughes and Westinghouse for ruby lasers and to American Optical for glass lasers.[28] Lacking a clearance, Snitzer could not work on the ARPA project,[29] but he was able to continue working on glass lasers on a separate unclassified project.[30] Project SEASIDE also supported a study on how laser pulses affected potential targets, essential in assessing how much damage lasers could do.[31]

PUBLIC IMAGE OF "THE INCREDIBLE LASER"

The pace of nonmilitary laser development was stepping up as Project SEASIDE was pumping up military spending. At Bell Labs, Alan D. White and J. Dane Rigden built a new type of neon laser that emitted a continuous red beam, catching the public eye and becoming the world's most widely used laser well into the 1980s.[32] At TRG, Steve Jacobs and Paul Rabinowitz had doggedly worked on a variation on the potassium vapor laser Gould had proposed, using cesium, another alkali metal. They finally got it to work in the wee hours of a Saturday morning in March and spent the rest of the weekend collecting data. They hadn't been able to reach Gould over the weekend, but they arrived in this office Monday morning, notebook in hands, to tell him. Gould knew before they said a word. "Well, I'll be damned. You made it work!" he said.[33]

A type of laser that would have tremendous impact around the world was invented in the fall of 1962. Light was first seen from semiconductors in the early 1900s but got little attention until the 1950s as semiconductor electronics matured. In the summer of 1962, MIT Lincoln Laboratory observed the brightest light from any semiconductor when they flowed electricity through a tiny chip of gallium arsenide called a diode. The light was in the infrared part of the spectrum, but it was so much brighter than ever before measured that it got the attention of other semiconductor physicists when the MIT team discussed it at a conference in New Hampshire.[34] On the train ride home to upstate New York, Robert N. Hall of General Electric Research Laboratories realized he might be able to use the effect to make a tiny chip of gallium arsenide into a laser by polishing its edges. He quickly succeeded and was soon followed by three other groups.[35] It would take years to perfect semiconductor laser technology, but it would bring revolutions to telecommunications, audio and video players, and eventually to the high-powered solid-state lasers you will read about in chapter 9.

The diode laser would have tremendous long-term impact. But what was catching the public eye was how well lasers could do things that had been very difficult.

At American Optics, Elias Snitzer and Charles Koester were developing methods to treat eye disease with light rather than by cutting into the eye with scalpels. One use was to treat diabetic retinopathy, a leading cause of blindness in which abnormal blood vessels spread across the retina, blocking vision. Eye surgeons had tried using light from bright mercury lamps to coagulate blood in the abnormal vessels to stop their spread, but their effect was limited. Pulses from a ruby laser worked better.[36] Laser photocoagulation has become a standard tool for this common cause of blindness. Stuart H. Loory cited a more dramatic use of an American Optical laser, removing a tumor inside the eye, in his 1962 article "The Incredible Laser."[37]

Loory's article is a fascinating insight into how the public saw lasers. He opened by breezily describing a laser punching a hole through a tiny diamond in a couple of billionths of a second and firing a laser pulse at the moon. He quoted Charles Townes as comparing the laser's role to that of the vacuum tube in electronics. Loory also expressed hope that the laser might have "even greater use as a peaceful tool than as a weapon." Yet he also wrote that 95 percent of the

$16 million that the US government had so far spent on laser research was for weapon-related research.

That was hardly surprising. The Cold War was at its peak, and Russian ships carrying nuclear missiles were still on their way home from the just resolved Cuban missile crisis. US Air Force chief of staff Curtis E. LeMay said energy from laser beams "could travel across space essentially with the speed of light. This would be an invaluable characteristic for the interception of ICBM weapons and their decoys."[38] If laser weapons lived up to their promise, a firm called Technology Markets Inc. envisioned a $1.25 billion market in 1970 for "huge lasers, not unlike the anti-aircraft searchlights of World War II, that would roam the heavens, seeking out incoming missiles and destroying or diverting them with the powerful beams of light."[39] Needless to say, nothing of the sort happened.

With that sort of hype, it was no wonder Schawlow made a point of calling his research lasers "credible." There weren't many of any kind of lasers available in 1962. Some engineers and scientists built their own lasers. Others bought ruby lasers from a handful of companies including Hughes, American Optical, the defense firm Raytheon, and TRG. Ted Maiman had founded a laser company called Korad Inc., short for coherent radiation, backed by an investment from Union Carbide. Trion Instruments in Ann Arbor was a spin-off of the University of Michigan. Small companies called Maser Optics and Optics Technology also made laboratory lasers. Spectra-Physics Inc. was formed to make helium-neon lasers in 1961.[40] None of them mass-produced lasers.

Most lasers were used in research and development. Some companies had big plans for lasers. Most notable were the AT&T plans to use laser beams to carry signals, through buried hollow tubes, for their PicturePhone video calling service, which Bell Labs was gearing up to show at the 1964 World's Fair in New York. Anything that used lasers was news.

BIG BANGS

Project SEASIDE's goal was to ramp up the power available from lasers by increasing the bulk of the laser. Glass looked good because it could be made in large chunks. Hughes scientists had learned how to accumulate large amounts of energy in a glass laser rod by changing the properties of the laser mirrors to stop it from lasing for a while, then switching the mirrors to release all the energy in a very rapid burst. The idea was that depositing all that energy at once would shatter a target like a nuclear warhead, like hitting it hard with a hammer.

However, when Westinghouse tried the process with ruby rods, the laser rods shattered. The same thing happened at American Optical with glass rods, which were thought more promising because they could be made much larger.[41]

The problem wasn't just a simple matter of thermal strain that can shatter a hot glass bowl put into cold water. It seemed fundamental. "The glass was beautiful from the point of view of optical quality; you could look through a long length like that and could see no striae, no ripple at all. [But] when you make laser rods of it, and you start to lase it, you blow it up every time," recalled Snitzer. "Apparently it was loaded with platinum, [because] it was made in a platinum crucible and made under reasonably oxidizing conditions, and that's the conditions under which you do get a fair amount of small platinum particles in the glass," he said. "It was just loaded with these little particles. Of course, you don't see them by looking at it, but the laser light sees it and it heats up the glass and it just blows up."[42] The problem was that the processing used to make the special laser glasses required melting the glass in platinum crucibles, which left tiny bits of metal embedded in the glass. Those tiny platinum particles absorbed light when the lamp or the laser was on, and the hot platinum expanded faster than the glass, shattering it.[43]

Bill Shiner, who began working at American Optical in 1962 while studying electronics in night school, helped Snitzer build a big glass laser. It was able to produce pulses to five thousand joules, a record at the time. The laser rod was about the size of a man's arm. But operating at those powers was dangerous, and the laser rod wasn't the only peril. "In Eli's laser lab we had two big metal waste-baskets. One said 'Eli' and one said 'Bill.' The reason was that the flashlamps used to blow up when you charged them, so we put [the metal trash cans] over our heads in case the lamp failed. We would hear the explosion and the glass would hit the metal wastebasket," Shiner recalled.[44]

Other problems also emerged, and by the July 1963 Project SEASIDE meeting in Woods Hole, Massachusetts, the power limitations in solid-state lasers had become quite obvious.[45] Two young officers from the Air Force Special

Weapons Command in Albuquerque said solid-state lasers were not viable for missile defense. Less than 1 percent of the electrical energy going into the power supply emerged in the laser beam, said Captain Donald Lamberson and Lieutenant John C. Rich. Generating a laser beam powerful enough to destroy nuclear warheads would require the electrical output of several Hoover Dams. Waste heat would build up in the glass and make it impossible to focus the beam. Lamberson recalled, "We showed conclusively that you really just couldn't get there. Energy levels were about four orders of magnitude too short; the precision of pointing was at least two orders of magnitude deficient; and even to get these inadequate capacities, you had to have all the glass in the world."[46] To make matters worse, the Soviets could easily overwhelm the laser defense by making warheads shiny and plentiful.

Figure 3.1. Elias Snitzer with a thick rod of neodymium-doped laser glass he used in laser experiments at American Optical. (Photo courtesy of the Snitzer family.)

By 1965 Project SEASIDE was limping to a close and laser weapons looked dead in the water. The problem wasn't just the lasers. After the light exited the laser, it had to travel through the atmosphere to hit the target. The air currents that make stars twinkle could bend the laser beam away from its target. The small fraction of the laser energy absorbed by the air would make the air at the very center of the beam expand and bend the beam like the heat rising above a hot parking lot on a sunny day. Bad weather would absorb much of the beam.[47]

Laser weapons had almost hit a dead end. Then a rocket scientist had a brainstorm.

THE ROCKET-ENGINE LASER

Arthur Kantrowitz, born in 1913 in New York City, had started out to be a physicist, but the Great Depression diverted him to fluid dynamics. Later he went back to complete a doctorate under Edward Teller at Columbia, then joined the Cornell University faculty.[48] In 1954 he had a fateful cocktail chat with a US Air Force officer and the president of the Avco Corporation about the difficulty of engineering nuclear warheads to withstand reentry into the atmosphere in order to strike their targets. Kantrowitz boasted that he could solve the problem in six months. The air force officer ventured that the Pentagon would pay to solve that problem, and Avco offered to support the lab. Thus was born the Avco Everett Research Laboratory at an old tire warehouse that Kantrowitz picked for its proximity to MIT, Harvard, and good sailing.[49]

His effort was a huge success, leading to the ablative coverings still used on space capsules returning to Earth, as well as on ballistic missiles.[50] It gave Kantrowitz a reputation as a wizard of the world of fluid and gas flow, important in a wide range of fields. Although the lab had a formal management structure, in practice everyone reported to Kantrowitz.[51]

After hearing a talk by Charles Townes, Kantrowitz wondered how he might be able to use gas dynamics to extract more power from a laser.[52] The answer came from the discovery of a new type of laser in 1963 by C. Kumar N. Patel, a young physicist at Bell Labs. Most previous lasers had emitted light when electrons changed their energy levels in atoms. Patel used carbon dioxide, a molecule that can emit light when its three atoms change how they vibrate or rotate. It worked quite nicely as a laser with pure carbon dioxide, emitting ten milliwatts. Adding nitrogen made a much better laser. "You put the two things together and the next day you get 10 watts from the same tube that gave you ten milliwatts before," Patel said. By mid-1965 he was up to two hundred watts, more power in a continuous beam than any other laser on record, and more than enough for anything Patel wanted to do at Bell Labs, so he moved on to other things.[53]

By then, however, Patel's laser had caught the eye of Ed Gerry, an MIT graduate student working part-time at Avco while finishing his thesis, who ran into Kantrowitz's office with a copy of Patel's first paper on the carbon dioxide laser in 1964. Avco had already been working on gas flow in lasers, but they had been stuck with a misconception that dated back to the 1930s. The three atoms in carbon dioxide molecules could vibrate in different ways, and physicists had always thought that the different vibrations died down at the same rate, like piano keys held down so the notes faded slowly. But Patel's experiment showed that the molecular vibrations faded at different rates, as if you took your fingers off some keys but not others. The fact that molecules kept vibrating much longer in one state than in others let Patel produce a large excess of molecules in that state so he could make a laser beam.[54]

Patel's carbon dioxide laser already had more power than other gas lasers that emitted a steady beam. Although the laser's output power was high, its wavelength was about ten micrometers, deep into the infrared spectrum, about twenty times longer than visible light. Fortunately air is reasonably clear at that wavelength, so the beam could travel far enough through air to be interesting for applications including laser weapons.

Kantrowitz decided to try his gas-flow idea on the carbon dioxide laser. Initially he thought of both zapping the gas with an electric charge, as Patel had, as well as using Avco's expansion technique, but on this new laser. But soon Gerry and others at Avco realized that expanding a hot gas through nozzles into a low-pressure region would produce a large enough population inversion to make a laser without using electricity. The same idea came independently to Nikolai Basov in the Soviet Union and Abraham Hertzberg at Cornell.[55]

Avco Everett was a defense contractor, so they were looking for lasers powerful enough to kill hard targets like ballistic missiles and nuclear weapons.

For continuous-output lasers like carbon dioxide, that meant a megawatt-class output that could be focused to deliver kilojoules of energy per square centimeter on the surface. "Lasers were not that efficient at the time, so you needed to get rid of a lot of waste heat. The best way to do that was to flow it away, the famous garbage disposal principle," says Gerry.[56] That is, expand the hot gas through a set of nozzles into a laser chamber, extract the laser beam out the side, and blow the hot gas out of the chamber, taking the heat with it.

The laser worked like a rocket engine, which might generate a gigawatt of raw power. "If you could capture one percent, or just a tenth of a percent of the energy in laser light, you were there" with the megawatt-class beam needed for a laser weapon, Gerry said.[57]

What Avco dubbed a gas-dynamic laser was simple. They burned a gas called cyanogen to produce a mixture of hot carbon dioxide and nitrogen,[58] which flowed through the system, emitting light after the chemical reaction. "It was a very simple thing, but not an efficient laser," says Gerry. That meant only a few tenths of a percent of the energy used to generate the flow was turned into laser light.[59]

Experiments started in a shock tube, then in a series of larger combustion-driven lasers. The next step was a small combustion-powered laser, then a larger one. They produced a good quality beam and record power of ten kilowatts after Gerry accidentally introduced some water vapor into the mixture of nitrogen and carbon dioxide.[60] By the time their rocket-engine lasers reached twenty kilowatts, they knew they could reach the power needed to use it as a weapon.[61]

Avco had its marketing manager at the time, Ben Bova, organize their first briefing of Pentagon officials on the breakthrough. It was an apt choice; Bova was a budding science fiction writer who had already published two novels.[62] The Pentagon had written off gas lasers, since they thought such lasers could not generate the needed energy, so they were stunned to discover that the Avco rocket-engine laser was more like a flamethrower than a flashlight.[63] Later, after publishing more novels and becoming editor of *Analog Science Fiction*, Bova jokingly told Boston-area science fiction fans that his job at Avco was to write a special type of science fiction for military audiences.[64]

The Pentagon classified the breakthrough, but ARPA and the Institute for Defense Analyses initially remained wary of the unfamiliar technology. It was part ignorance and part "not invented here," the preference of large organizations for using their ideas. The agencies asked other contractors to scale up electrically powered carbon dioxide lasers based on Patel's design. Those lasers converted more of the input power into light, but had no way to get rid of the waste heat, and became behemoths. Hughes needed fifty-four feet of tubing to generate 1.5 kilowatts for ARPA.[65] Raytheon generated 8.8 kilowatts and claimed 13 percent efficiency but required about six hundred feet of tubing.[66] At those sizes and power levels, they couldn't compete with the rocket-engine laser.

Progress with rocket-engine lasers soon spoke for itself. Avco cranked the power up to an impressive 138 kilowatts with its MK-5 system in March 1968. United Aircraft's Pratt & Whitney division also got into the race, trying to win business from the air force, and built at its new test facility in West Palm Beach, Florida, a laser called XLD-1 that reached seventy-seven kilowatts in April. Avco gradually wore down Pentagon committees sent to investigate, and their results finally changed minds. "The change from the Pentagon refusing and resisting to fund anything to everybody wanting to fund it took place about 5 o'clock one evening," Kantrowitz said. "Suddenly all the [armed] services wanted to get us under contract."[67]

EIGHTH CARD

In 1968, ARPA launched a hush-hush classified project to scale up the gas-dynamic laser. It was called Eighth Card, an allusion to the advantage an eighth card could give in seven-card stud poker.[68] It was so secret that the only way to get information on the project was to be invited to work on it, even if you had a secret clearance.[69] Even the term "gas-dynamic laser" was highly guarded, and nobody in the military called it a rocket-engine laser.[70] The Air Force Weapons Laboratory in New Mexico also sponsored other rocket-engine laser development at both Avco and at Pratt & Whitney in Florida.[71]

The two programs soon diverged. ARPA continued to focus on defense against Soviet nuclear ballistic missiles. Keenly aware that Basov was working on gas-dynamic lasers and had high political connections in Moscow,[72] they worried about a Soviet breakthrough in laser weapons.

The weapons lab took the lead for the three armed services in studying prospects for battlefield laser weapons. So each service could carry out its own tests, the lab spent $5.8 million on the Triservice lasers, three 150-kilowatt rocket-engine lasers. Avco was to deliver one to the weapons lab; a second to the Army Redstone Arsenal in Huntsville, Alabama; and a third to the Naval Research Laboratory in Washington, DC.[73] Avco also built huge flowing-gas carbon dioxide lasers powered by lightning-like bolts of electricity, with names like Thumper and Humdinger. Thumper was the size of a house, drew power through a bank of electrical cables as thick as a person's arm, and got its name from the massive THUMP it made when fired.[74]

Although the Pentagon long kept the nature and goals of the Eighth Card project highly classified, they declassified the rocket-engine laser concept after Basov openly published about it in Soviet journals.[75] They allowed Gerry to disclose the new concept at an April 1970 conference of the American Physical Society, and follow that with a detailed article in *IEEE Spectrum*.[76] But Gerry could not mention that power had exceeded fifty kilowatts, although Avco had already passed one hundred kilowatts. Versions of the graphs that Gerry presented at the meeting were cut off abruptly at the maximum power level that security rules would let him discuss in public.[77] At a physical society press conference, Gerry compared the beam's power to that of a small car engine, and he was surprised the next morning to find a headline reading, "Laser Powers Small Car."[78]

A SOVIET LASER HANDGUN

In the same time frame, engineers at the Peter the Great Academy for Soviet missile specialists were developing a real laser handgun intended for cosmonauts who were to be stationed on the Almaz Soviet military space station, eventually launched in 1973. That project was highly classified and concealed as part of the civilian Salyut series of satellites.[79] The laser pistol was a souped-up version of the early flash-powered lasers demonstrated by Maiman and Snitzer. However, light came not from flashlamps but from metallic flares that burned much more intensely to fire a single laser pulse lasting a few milliseconds. The laser gun would hold eight packaged flares, ejecting each one after it was used, like bullet casings.

The laser gun apparently was intended to defend the military station against US piloted or robotic spacecraft that the Soviets thought might try to capture some of their classified equipment. It wouldn't destroy the approaching spacecraft, but the brilliant flash would knock out sensors or vision, perhaps permanently. However, it was never used, because it fell far behind schedule and was not ready until 1984, long after the last Almaz satellite flew.[80]

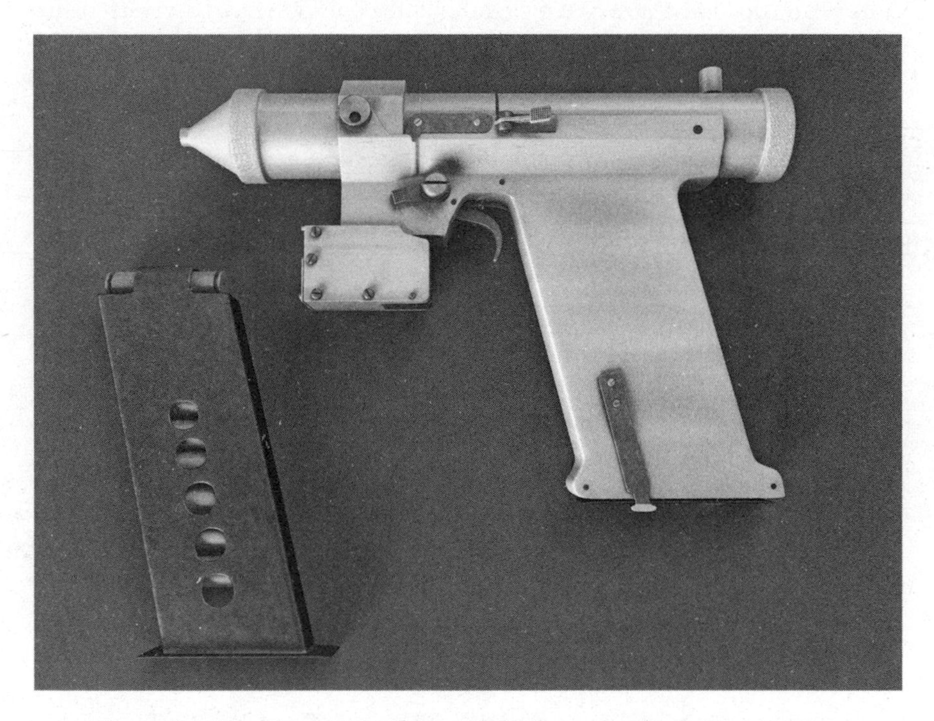

Figure 3.2. Soviet engineers designed this laser pistol so cosmonauts on military satellites could defend themselves from astronauts or robotic satellites trying to steal Soviet secrets. It never got into space. (Photo courtesy of Anatoly Zak / RussianSpaceWeb.com.)

THE TRISERVICE LASER

The Triservice Laser was the sort of bold idea that got people into trouble. Avco had optimistically thought it would be easy to get more power by scaling up their earlier rocket-engine laser. But once they started, they had to go back to basic physics to work out design problems. It took a year before the first parts reached the weapons lab in April 1970, and it took another year and a half to get the laser to the promised power. Even then, the beam was so diffuse it was more like a large searchlight emitting invisible heat rays than a deadly laser weapon.[81]

Lamberson, who had earned a PhD in aerospace engineering after pointing out the limits of solid-state lasers, took over the weapons lab's laser division in mid-1970 as a lieutenant colonel. He was bright, energetic, and charismatic, and he eventually ran out of patience with Avco. In December 1971 he pushed to take over the project, and within a year the weapons lab had greatly improved the beam. They started taking laser pot shots at slow-moving drones in mid-1973. The first dramatic flash they saw came from scorching some metal supports on a water tower, which survived. When they improved their aim, they managed to burn through the skin of the drone, causing it to crash but with only minor damage. On the next day, November 14, 1973, they managed to keep the laser focused on a fuel tank for a little over a second, enough to heat fumes in the tank and trigger a satisfying explosion.[82]

Details were kept classified for years, but I saw a film of the test at an April 1982 laser conference. The infrared laser beam slicing through the air was invisible, but I could see the drone and see a spot on it start to glow as the beam heated it. The film was shot from a distance, and its quality wasn't great, but I wrote shortly afterward that the film "was adequate to show the drone flying through the air, being illuminated by the laser, and eventually catching fire as the fuel tank ruptured. In a second test the laser cut the control wires and caused the drone to crash out of control."[83] It was a long way from the instant death of a science fiction death ray, but it showed that the rocket-engine laser might have weapon potential. That made it an important milestone in convincing military brass to keep supporting laser weapon development.[84] But much more remained to be done.

MORE THAN JUST LASERS

Shooting down the drone made the rocket-engine laser more of a weapon than anything "death ray" Matthews had ever made. It was the equivalent of killing the sheep the British Air Ministry had set out for would-be death-ray demonstrators. The demonstration was a milestone for speed-of-light weapons, but much more work remained, and it wasn't just building bigger and better lasers.

The big question for any weapon system is, How lethal can it be in the real world? It's one thing to shoot a slow-moving drone that might be only a few hundred feet away on a test range. But it's far more difficult to hit targets that move faster, are farther away, and are harder to destroy. The question of lethality depends on more than just the laser. It also depends on the optics that focus the laser beam through the air, how well the air transmits the beam, how well the beam can be aimed so it can stay focused on the target long enough to destroy it, and how the beam interacts with the target. Moreover, the measurements must be extremely precise and extremely reliable.

To start at the beginning, the laser beam has to exit the laser and be directed toward the target. For rocket-engine lasers, this meant going through some kind of portal that separated the hot, thin gas emitting light inside the laser from the denser air outside. Initially rocket-engine lasers had no solid windows, only aerodynamic windows created by gases flowing at supersonic speed past a hole in the wall of the laser. The gas moved so quickly past the hole that it didn't have time to leak out, and produced a shock wave that confined the gas. That transmitted the laser beam surprisingly well, but it couldn't focus light. That required the laser weapon program to develop special focusing optics that could withstand high light intensities and deliver lethal energy to targets.[85]

That wasn't all. We don't notice a little light absorption in ordinary life, like the few percent of light absorbed by window glass. But if you focused a megawatt laser beam through a window, a few percent of loss would add up to tens of kilowatts and a melted or shattered window. Mirrors can be easily made to absorb less light than lenses, so they usually are used for focusing laser beams, but it's still very important to minimize absorption so little power is lost. It's also important to make the surfaces extremely smooth to avoid scattering laser light so it's lost from the laser beam. Thus special high-performance optics had to be developed for the rocket-engine lasers emitting at ten micrometers, a wavelength where few good optical materials were available.

This was a job that needed world-class scientists who specialized in esoteric areas of optics, like Jean Bennett at the Naval Weapons Center in China Lake, California. She earned a doctorate studying spectroscopy, the light absorbed or emitted by atoms and molecules, and in the process became an expert in making and measuring optics. She was an avid outdoorswoman and enjoyed hiking the area around China Lake with her family.[86]

Bennett was absolutely meticulous in her measurements, with the kind of care and precision needed for pushing the cutting edge of the technology. She used that care and precision to film family trips, producing sixteen-millimeter films that looked like National Geographic specials. That precision helped her understand and learn how to control tiny imperfections in optical surfaces that could make sure megawatt-class laser beams wound up on their targets instead of melting the mirrors.[87]

Another crucial question was how well the beam could travel through air. We often think the air is clear, but that's because our eyes evolved to see the light that reaches the ground. Yet the air absorbs many other wavelengths. It absorbs most of the sun's ultraviolet light, particularly the shortest wavelengths that are the most likely to cause skin cancer. The air also blocks much of the infrared light but has some clear windows. One of those windows is at the ten-micrometer wavelength of the carbon dioxide rocket-engine laser, but not all of the laser light gets through the air. In fact, even at visible wavelengths, air currents and turbulence can make distant objects look wavy. Look across a hot parking lot on a sunny day, and you can see currents of hot air rising and bending light. Haze, fog, rain, or snow can blur your view.

Bell Labs engineers saw these problems when they hauled one of the first ruby lasers to the top of a microwave tower one night and fired pulses at a rooftop twenty-five miles away where another engineer stood with a movie screen, watching for red pulses. He didn't see many.[88] Military researchers had seen similar effects with ruby and glass lasers, and carbon dioxide in the air made the effects worse for lasers emitting at ten micrometers. They had to find out if a powerful laser beam could blast holes through that absorption, or would it just heat the air and make it worse?

The toughest question of all was the bottom line: would the laser beam disable or destroy the target? That depended not only on the quality of the optics and the clarity of the air but also on how the beam interacts with the surface, and those effects can be subtle.

Science fiction blasters fire bullets of energy that travel in a straight line and explode on contact in a fireball. Bullets and missiles either miss or stagger targets with a kinetic impact that damages or destroys them. Real-world laser weapons focus on a spot on the target and heat it, like sunlight focused by a lens. Heating a target requires a bright laser beam, optics that can keep the beam tightly focused onto the target spot, and a target that absorbs light at the laser wavelength. Online videos of laser weapons show the laser spot appear on the target and grow steadily brighter as the heat builds up until the target reaches the melting or ignition point.[89]

The effect of heat delivered over a matter of seconds depends on the nature of the target. Focusing a laser on a stationary target that is not highly flammable may not work well because smoke and other debris from heating the target may rise to block the beam from heating the target further. On the other hand, the smoke blows away if the target is moving and the laser can track it accurately to continue the heating. A target loaded with explosives can be destroyed by heating its shell without penetrating the skin, until the heat ignites the explosives and produces high pressure that explodes the shell. But burning through the skin of a drone to disable its flight may not ignite the explosives, which could survive to detonate when the drone hits the ground, potentially harming people the laser was intended to protect. The Airborne Laser, described in chapter 6, was designed to kill boosting long-range ballistic missiles by structurally weakening an internal tank so it unzips and their payload falls down a short distance from the launch site.[90]

Laser weapons intended to deliver short pulses to crack open nuclear reentry vehicles or other brittle targets would be hampered by another lethality issue. High laser intensities can ionize the air to produce a plasma, and this kind of ionization effect could block transmission of the laser pulse before it delivered enough energy to damage the target. This could have been added to Lamberson's list of problems with pulsed solid-state lasers in the mid-1960s.[91]

BETTER LASERS THROUGH CHEMISTRY

Gerry's 1970 article on gas-dynamic lasers mentioned that using different chemicals in rocket-engine lasers could produce different laser wavelengths.[92] The most important of these alternative chemistries proved to be combining compounds to produce excited hydrogen fluoride molecules that when flowed through a rocket-engine laser cavity could emit at wavelengths less than the ten-micrometer wavelength of carbon dioxide.

The first such chemically powered laser was demonstrated in 1965, but it did not use flowing gas.[93] In 1969, researchers at the Aerospace Corporation in Los Angeles used a rocket-engine laser design to generate up to 630 watts of infrared light, amounting to about 12 percent of the energy from the chemical reaction.[94] The chemistry was potentially nasty, and the resulting hydrogen fluoride was toxic and needed to be captured and detoxified, but it promised to be more compact than the gas-dynamic carbon-dioxide rocket-engine laser as well as more powerful, which attracted military interest.[95]

One issue was that hydrogen fluoride lasers normally emit at 2.6 to 2.9 micrometers, which air absorbs strongly. Fortunately, replacing the common isotope hydrogen-1 with the heavier deuterium (hydrogen-2) shifts the wavelength to around 3.8 micrometers, where air is more transparent.[96] Deuterium is costly enough that most experiments used hydrogen-1 to save money, but air transmitted the deuterium wavelength well enough to make it attractive as a weapon.

Hydrogen fluoride offered higher power from a rocket-engine laser than what was available from carbon dioxide. Its shorter wavelength was also important because the size of the optics needed to focus the beam increased with wavelength. So a 3.6-meter mirror could focus a hydrogen fluoride beam onto the same size spot as a ten-meter mirror could focus a carbon dioxide beam. Better yet, the hydrogen fluoride laser mirror would cover only 13 percent as much area, making it much lighter.

Chemical rocket-engine lasers had reached the kilowatt level when Ed Gerry moved to ARPA in 1971 to run its laser weapon program. With rocket-engine and discharge-powered carbon dioxide lasers having reached high powers, ARPA turned them over to the armed services. He supported the Baseline Demonstration Laser, the first high-energy chemical rocket-engine laser,[97] which exceeded a hundred kilowatts in 1973, leading the way to higher-power chemical lasers.[98] He also looked for other high-energy lasers that might make better weapons.

Rounding up the state of laser progress in mid-1972, *Aviation Week* reporter Philip J. Klass wrote, "Most experts in the field agree that [continuous] power levels of at least a few megawatts should be available by the end of the present decade. There is more disagreement over the question of how big an impact such devices will have in military weapons, strategy and tactics." Although some "enthusiasts" expected laser weapons to eventually be able to block aircraft and missile attacks, he wrote, "Even such visionaries agree that this time is at least several decades away."[99]

THREE SERVICES GO IN THREE DIRECTIONS

The three armed services took different directions on laser weapons, reflecting their different needs.

The army, which needs to operate over all terrain, shoehorned an electrical carbon dioxide laser into a tanklike vehicle and shot down drones but decided it would be impractical on the battlefield and largely abandoned laser weapons.[100]

Navy battleships can accommodate much larger equipment, but they live in a marine environment, where water vapor absorbs light from carbon dioxide lasers. The navy hoped marine air would be clearer at deuterium fluoride wavelengths, so it turned to chemical lasers. After ARPA's Baseline Demonstration Laser test, the navy teamed with the agency to build the Navy Advanced Chemical Laser, which reached four hundred kilowatts with deuterium fluoride and shot down a couple of small missiles.[101] That laser was a monster, occupying a group of small buildings at a test site operated by the aerospace firm TRW Inc., in the hills east of San Juan Capistrano, California.[102]

Then the navy asked TRW to build the Mid-InfraRed Advanced Chemical Laser (dubbed MIRACL). The rocket-engine laser was a maze of plumbing that mixed gases to produce chemical reactions yielding deuterium fluoride that was fed into the Sea Lite Beam Director, with a 1.6-meter mirror (see photo insert).

Officially, it was the first megawatt-class continuous-beam laser when it was completed in 1980.[103] Unofficially, the rocket-engine laser reached a record 2.2 megawatts.[104] However, by the time MIRACL was up and running in the early 1980s, the navy had found the deuterium fluoride wavelength was absorbed by moist marine air and lost interest in the giant laser. After Congress dropped the navy project funding from the federal budget, MIRACL became the centerpiece of the High Energy Laser Systems Test Facility at the White Sands Missile Range in New Mexico, where it was used for military tests, including shooting down rockets and illuminated orbiting satellites.[105] Over more than fifteen years of operation, it was used in more than 150 tests up to seventy seconds each, but altogether amounting to less than one hour of total lasing time.[106]

THE QUEST FOR OTHER LASERS

Meanwhile ARPA focused on finding new types of laser, preferably at shorter wavelengths that would require smaller optics and be little absorbed in air.

One idea was zapping normally inert rare gases like xenon, krypton, and argon with high-voltage electrical pulses. The original idea was to blast off electrons so pairs of rare gas atoms would form short-lived molecules that emitted ultraviolet light when they split apart. Laboratories in the United States and Soviet Union demonstrated such lasers in 1972 and 1973.[107]

But the real payoff came when later J. J. Ewing and Charles Brau at Avco started looking at mixtures of rare gases with the halogens fluorine, chlorine, bromine, and iodine. Another researcher had observed ultraviolet light when chlorine reacted with xenon, so Ewing and Brau analyzed the physics.[108] Then they decided to zap a mixture of xenon and iodine with an electron beam to see if it would lase. "A tremendous amount of light came out," Ewing said. The wavelength was close to what they had predicted, and the light was so bright he wondered, "What have we been doing wrong all this time?"[109]

They never did make a xenon-iodine laser, but after a team at the Naval Research Laboratory made a laser from a similar compound, Avco rearranged their experiment to crank up the electron-beam power, and in June 1975 they started blasting away. First they made xenon fluoride lase, then xenon chloride.

Then they tried krypton fluoride, which they had been wary of, and got by far the brightest output of any of the lasers.[110] They had found a whole family of bright new ultraviolet lasers. The lasers only fired brief pulses, but they produced much more energy than any other ultraviolet laser. ARPA had a hot new weapon candidate.

THE AIR FORCE GETS THE AIRBORNE LASER LABORATORY OFF THE GROUND

The Air Force Scientific Advisory Board pondered the potential of laser weapons for much of 1969 and 1970. In the spring of 1971, the group met at Kirtland Air Force Base, the home of the Air Force Weapons Laboratory. With optimism growing about prospects for laser technology, the board concentrated its discussions on what it saw as a young technology with exciting prospects. Powers had reached the one-hundred-kilowatt range, reviving the excitement about prospects for laser weapons.

Weapons were not the only frontier for high-energy lasers. The latest idea in nuclear fusion was inertial confinement, using intense laser pulses to heat and compress a pellet loaded with hydrogen isotopes to trigger nuclear fusion. Ostensibly, the Atomic Energy Commission was pursuing the idea for fusion energy, but behind the scenes the real interest was in simulating the explosions of thermonuclear bombs to better understand their physics.[111] Interest was growing in a separate energy-related laser concept, using lasers to increase the concentration of the fissionable isotopes of uranium. Researchers at the Avco Everett Research Laboratory had their own process. The public rationale was to produce fuel for nuclear reactors, but hidden in the background was interest in refining the plutonium used in nuclear weapons by removing isotopes that might dampen the explosion.[112]

One of the air force science advisors, William McMillan of UCLA, suggested creating a national laboratory for laser defense modeled after the Manhattan Project.[113] In an opening talk Edward Teller urged the air force to "pay early attention to lasers and the changes that the existence of this instrument could bring about." Later he got up and walked around, gesticulating, envisioning an undefeatable "great laser battleship in the sky," and proposing a laser development program for the 1970s comparable in scale to the US development of long-range nuclear missiles in the 1950s. Others lauded prospects for developing powerful rocket-engine and electric-discharge lasers.[114]

Lamberson, normally a great fan of lasers, wound up as a voice of moderation, warning that any laser program should be a systematic one based on sound technology. He opposed "Manhattan or multi-hundred megabuck programs" until the technology was ready. He spoke partly from his knowledge of the state of laser development and partly from concern that such plans would take the laser program away from the weapons lab.[115]

In the end, the panel favored Lamberson's moderation and recommended a more modest project that became the Airborne Laser Laboratory at the weapons lab.[116] The roots of that project trace back to a 1967 discussion between two officers at the weapons lab with very different backgrounds. Lieutenant Colonel Howard W. Leaf was a fighter pilot who had moved up the chain of command to "fly a desk" in the research bureaucracy. First Lieutenant Petras V. Avizonis was a Lithuanian-born PhD in physical chemistry with an interest in lasers inspired by a talk by Charles Townes. Leaf wanted to give pilots more firepower, particularly on otherwise defenseless Airborne Warning and Control System (AWACS) radar planes, and thought that "bits and pieces" of laser technology being developed at the weapons lab could be demonstrated on an airborne test bed. Avizonis liked the idea, and it became part of an advanced development plan.[117]

Nothing happened at first, but the air force did put the weapons lab in charge of its laser weapons program.[118] Lamberson took over the weapons lab laser program in March 1969.[119] He brought an enthusiasm for lasers and a willingness to tackle a tough technological challenge.[120] He also brought a range of talents. He knew the science, knew the technology very well, and also was a very good manager and very good leader.[121] He also had a gift for finding very bright people and helping them work together. He would need all those skills.

Lamberson had already gotten the Triservice Laser up and running. Like other rocket-engine lasers, it was a complex array of pipes, valves, and nozzles feeding into the low-pressure cavity where the beam was generated. Yet it was set firmly on the ground in a laboratory, with beam-directing optics tacked on.

The Airborne Laser Laboratory was also a test bed, but its goal was much more ambitious: a system-level test of the feasibility of building a laser weapon into an airplane. That meant the laser and all the equipment needed to zap targets had to operate in a constantly vibrating environment.

The first step was to prepare a military version of a Boeing 707 airliner to house a carbon dioxide rocket-engine laser emitting about four hundred kilowatts, powerful enough to be lethal and small enough to fit into the plane. It also required a fire-control system to identify and track targets, and optics to focus the beam onto the target. They couldn't use glass optics; they needed materials that reflected or transmitted at the laser's ten-micrometer wavelength. They also had to install a movable turret to point the beam at remote targets. And it all had to work together.[122]

The modified plane was delivered at the start of 1973, and tests of the optical and aircraft systems began with a lower-power laser emitting at the same wavelength, as Pratt & Whitney started building a laser that could fit into the plane. It took years to squeeze the laser into the plane, where the laser occupied the front third of the interior behind the pilot; the power supply and fuel occupied the middle, and the rear third held 1970s vintage computers and control systems.[123] Customizing the plane to fit the laser required cutting holes and changing it so much that people worried if it could fly properly. Odd vibrations plagued the plane on early flights. Everything had to be assessed and adjusted to fix what was misbehaving.[124] The air force got the laser-laden plane off the ground in 1978 for a photo, but it still needed work.

The seasoned traveler takes aircraft vibrations for granted. Optical systems don't. They include sensitive machinery that must stay carefully aligned for the laser to work properly. Constant vibrations can work things loose, and planes vibrate constantly in the air. The vibrations made the laser beam jitter and drift off target.

Getting the laser beam through the air proved to be a problem. Designers had chosen the carbon dioxide version of the rocket-engine laser because it was the most powerful type available in the early 1970s. But its ten-micrometer carbon dioxide wavelength did not travel well. Water vapor absorbed it and so did carbon dioxide molecules in the air, and that absorption heated the air, bending or dispersing the beam.

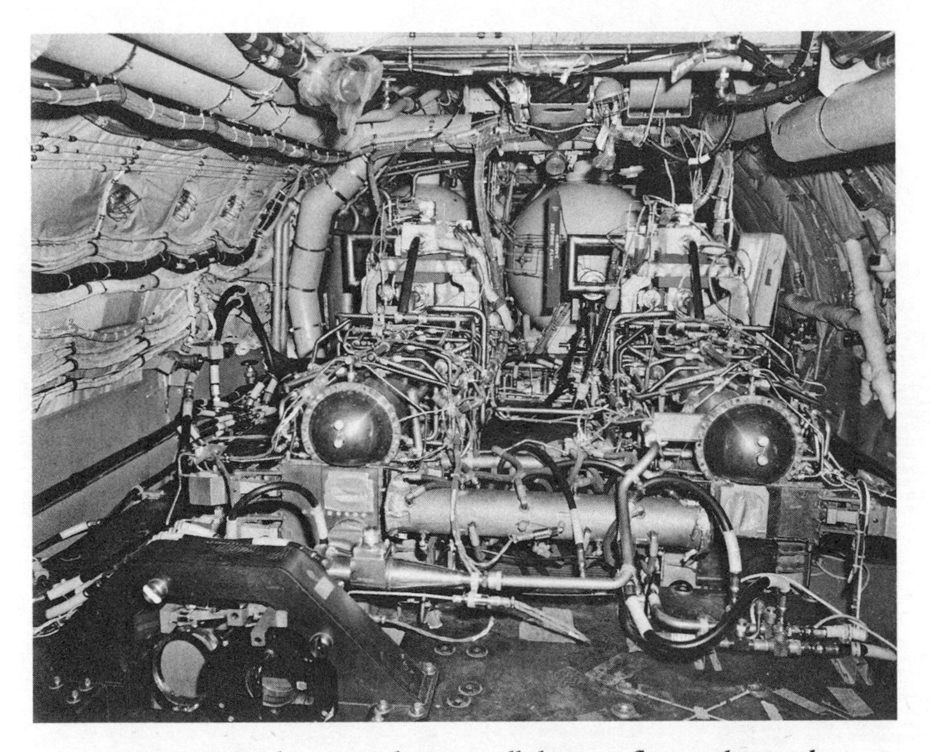

Figure 3.3. Rocket-engine lasers are all about gas flow, so they need lots of piping and valves. This is the interior of the Airborne Laser Laboratory, which flew in a military counterpart of a 707. Fuel tanks are in back. (Photo courtesy of the National Archives.)

The optics could only focus light onto targets within a few kilometers of the plane, too short a distance for many potential missions.[125] The metal mirrors absorbed so much heat from the high-powered beam that they had to be cooled with flowing water.[126] The powerful laser beam ignited dust particles, producing "fireflies" in the air, tracing the path of the invisible beam. The beam also burnt dust on optical surfaces, heating and damaging them, so some parts had to be assembled and serviced in clean rooms.[127]

The iron laws of the military bureaucracy also forced a change of management as crucial tests were coming. After an exceptionally long nine years on the job, Lamberson was promoted to brigadier general on February 2, 1978, and had to move on because the head of the laser program had to be a colonel. He had grown the job from a group of twenty-five people with a $6.5 million budget to a team of 350 and a budget of $86.8 million.[128] The challenge was to shoot down moving targets from the air.

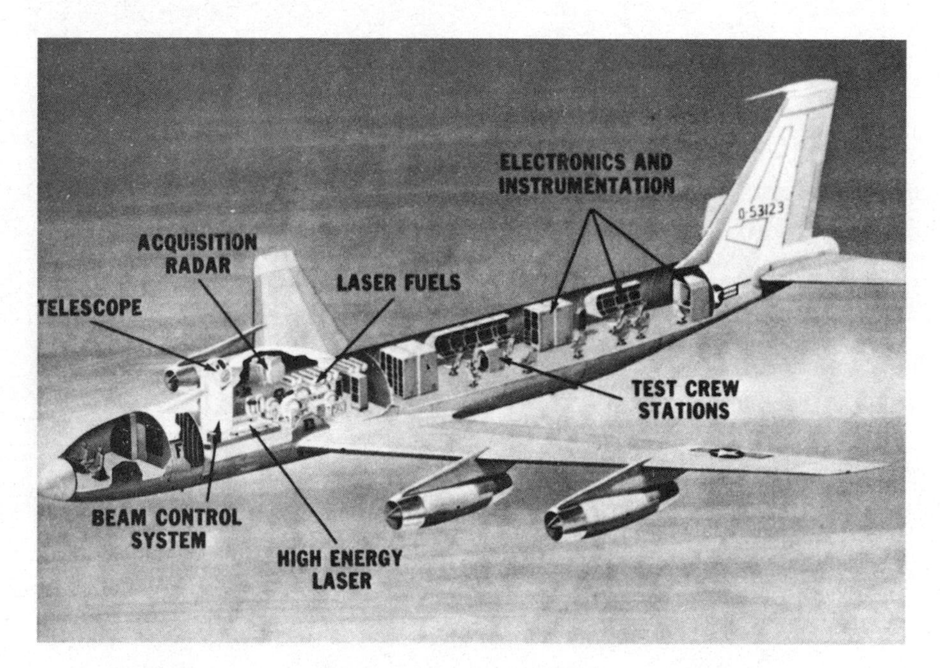

Figure 3.4. Layout of the Airborne Laser Laboratory. The laser itself, shown in figure 3.3, is marked as High-Energy Laser and occupies only a fraction of the cargo space. (Courtesy of the National Archives.)

On September 4, 1980, the weapons lab reported that all vital hardware worked, and the laser could produce light, although it had not yet emitted a beam. Some issues remained. Running gas through the laser produced thrust that pushed the aircraft's nose up at high altitudes, but the pilot could compensate. The cooling system ran a bit hotter than desired. The big question left was if optical-system vibrations would throw off beam alignment, but that could only be answered by going ahead with the next round of tests to see if the Airborne Laser Laboratory could shoot down a missile and a drone.[129]

The pressure was high. The program itself would be in danger if the laser couldn't bag a target soon. A round of ground tests found leaks in the cooling system, so equipment had to be removed, repaired, and reinstalled. Ronald Reagan's defeat of Jimmy Carter in November added to the time pressure. That meant that Hans Mark, a staunch advocate of the Airborne Laser Laboratory, would be out of his job as air force secretary when the new administration arrived in January. He and another program advocate, Senator Harrison Schmitt, were scheduled for a visit on January 15, so the weapons lab rushed to show them the laser firing a high-powered beam from inside the plane.[130] For over a week in early January, three shifts of workers worked around the clock to set up the laser.[131]

The weapons lab planned two test firings the day before the planned visit, but both had to be stopped prematurely because of gas-flow problems. The second fired a beam from the laser into the open air, an important first. But it also damaged a critical mirror that could not be replaced in time to repeat the demonstration the next day. Both Mark and Schmitt understood that the previous day's experiment had been a partial success, but the weapons lab team was disappointed.[132]

Another round of private tests followed, and a crucial demonstration was scheduled for late May 1981 at Edwards Air Force Base in California. Their goal was to shoot down a missile with the plane in the air. That required keeping the beam focused for two to four seconds on the nose of a missile three to five kilometers away. Although the laser could generate four hundred kilowatts, only about seventy-five kilowatts were expected to reach the target, making it harder to destroy the target. Officials were optimistic, but everything had to go just right.

Bad weather spoiled results on the first of three days of planned tests. On the second day, the beam briefly hit the target when the laser ran for 1.2 seconds. On the final day, the laser locked on the missile for the full 1.8 seconds that it ran, but it did not stay long enough on the right spot to report a "kill."[133]

Had the tests been conducted in secret, the weapons lab would have quietly noted the lessons learned and come back to try another day. But this test was public, and the press reported that the laser had flunked the test.[134] The air force was annoyed, but they shouldn't have been surprised.[135] Movie death rays always worked for the good guys, but in the real world Murphy's Law held that something would always go wrong. I called a colonel I knew at the Pentagon in mid-1981 to ask what was next, and he said I wouldn't hear anything until the Airborne Laser Laboratory shot something down.

That would take two years. Pentagon brass minced no words with Airborne Laser Laboratory program manager Colonel Jerry Janicke after he briefed them before planned tests at Edwards Air Force Base in May 1983: "Don't embarrass us, and get it done."[136] After a series of test shots, Airborne Laser Laboratory shot down its first two missiles on May 26.[137] They could finally report success.

LASERS EVOLVE BEYOND DEATH RAYS

The civilian laser industry also had come a long way since the days of the incredible laser. I started working for a laser trade magazine called *Laser Focus* in 1974. When I told a neighbor about my job a few months later, he immediately asked if I was writing about death rays. I explained that lasers were used for many peaceful applications and that we rarely covered laser weapons.

The magazine's main focus was on scientific research and industrial development. One hot research topic was the use of lasers that could be tuned across the spectrum to study the properties of atoms and molecules. A company called KMS Fusion was trying to develop laser-powered nuclear fusion as an energy source. Government laboratories were using lasers to try to separate isotopes. Laboratories were developing new types of lasers for specialized tasks.

Industry also was developing laser technology. We wrote about companies trying to develop video players that used lasers to play back movies recorded on large disks the size of phonograph records. We described how lasers could read new bar codes being developed for automated checkout in supermarkets. Companies were just starting to experiment with sending laser light through optical fibers for communications, and within a few years they would start testing the fiber-optic links in telephone networks. The cover of one issue showed a laser welder being tested in a Ford production line, and a news story on laser machining mentioned an Avco flowing-gas laser producing ten kilowatts of light, a little sibling of the lasers Avco was building for military customers. We also wrote about development of lasers for medicine.

Looking back, I find a few stories did relate to laser weapons. The cover of the

December 1974 issue showed a mirror that could adjust its shape to change how it reflected light. That was developed at the Itek Corporation, and one potential use was shaping the beams from high-powered lasers so they could focus better.[138] That technology was too immature for use on the Airborne Laser Laboratory, but eventually it would find use on a later giant flying laser covered in chapter 6.

Military labs also found peaceful uses for lasers that benefitted the world in other ways. In 1964, Emmett Leith and Juris Upatnieks, young researchers at a military lab run by the University of Michigan, displayed three-dimensional laser images called holograms that did not require special viewers. For a while holography became an art form.

Laser experiments rapidly grew more sophisticated. Not long after the ruby laser was invented, researchers tried bouncing laser beams off the moon but at best saw only dim red spots. Apollo astronauts Neil Armstrong and Buzz Aldrin took to the moon a special mirror called a retroreflector that could bounce light directly back to the point where it originated. The goal was to measure the distance to the moon more accurately than ever before.

Hal Walker, the young boy fascinated by a toy Buck Rogers ray gun, grew to be six feet four and in the early 1960s went to work for a laser company called Korad, which was founded by Ted Maiman.[139] Walker started as a technician and worked his way up to be manufacturing manager. One of his jobs along the way was aiming laser beams at the moon to bounce light off the retroreflector that the astronauts had left there. NASA wanted Korad to send its best man to operate a Korad ruby laser at Lick Observatory.

Walker had more adventures than he expected in getting Korad's laser up and running in the mountaintop observatory. Once the astronauts headed home, he had to scan the lunar surface to find the retroreflector and get a return signal. The laser beam spread across at least a mile on the moon, reducing the reflected power to a barely detectable level, so he had to increase the laser power, taking care not to damage the laser. The astronomers were surprised to find red spots from another laser appearing on the moon, so they had competition in the search. The laser heated up, so Walker asked for fans to cool the drive circuits. Finally the laser hit the retroreflector, sending a bright pulse back to the telescope, giving enough light to measure how long the laser pulse took to make the round-trip, and giving the first direct and precise measure of the lunar distance.[140]

The only laser weapon Walker worked on at Korad was fictional and came from the company being based in Santa Monica, just a short drive from Hollywood. The producers of *The Andromeda Strain* wanted to use lasers in the film, but Korad didn't want lasers to be seen just as weapons. They compromised by showing lasers being used in biological experiments but also being able to deal with the dangerous pathogen getting loose. They took a Korad laser to the studio and set up a smoke background to show the green beam. In the scene, a man infected with the pathogen was trying to escape up the ladder and turned around to face his pursuers. At this point, Walker's crew from Korad moved the actor out of the way, put a dummy in his place, and the film rolled as they zapped him with a big splash of laser light. It was all good fun for Walker and friends, but the management at Union Carbide, which by then owned Korad, had a collective cow about the public relations impact of the company name appearing in a scene with a laser death ray.[141]

A SHIFT IN FOCUS

On September 26, 1983, the Airborne Laser Laboratory shot down its final target. The laser beam hit its drone target three times in a row, but only on its final shot did it do enough damage to prevent the drone from completing its mission. In his book *Airborne Laser: Bullets of Light*, air force historian Robert Duffner wrote, "The final shoot-down still was an extremely important first step in proving that it was scientifically feasible to engineer and use a complex laser system to destroy aerial targets."[142] The shoot-down was a proof of principle. If everything went just right, the Airborne Laser Laboratory could shoot down a relatively vulnerable target.

However, that final test was far from a demonstration of a practical weapon system. The Airborne Laser Laboratory was not the ultimate weapon for the battlefield. It was too bulky, too inefficient, too finicky, and too maintenance-intensive to be a practical weapon. Even when it worked, it could not reliably deliver the full power of the beam onto a target. The air absorbed too much light at the carbon dioxide laser's ten-micrometer wavelength, not only reducing the power transmitted but also deflecting much of the light away from the target. The Airborne Laser Laboratory had generated a long list of valuable lessons learned.

One of the most important of those lessons was that rocket-engine lasers were not ready to meet military requirements in the battlefield. The army had reached that point first, and the navy had followed after its struggles with MIRACL. Now the air force decided to turn away from developing battlefield lasers. They "orderly terminated" the program, gradually winding it down to keep the Airborne Laser Laboratory available for further tests if needed for a few years. In the end, nobody wanted it enough to pay the $85–$148 million that would be required to get it ready for further tests. On May 4, 1988, now major general Lamberson and hundreds of others came to watch the Airborne Laser Laboratory take off from New Mexico to fly to the Air Force Museum in Dayton, Ohio.[143] It is now in storage.[144]

Figure 3.5. Donald Lamberson, the driving force behind the Airborne Laser Laboratory as a major general during the 1980s. (Photo courtesy of AIP Emilio Segrè Visual Archives, Hecht Collection.)

The Pentagon's focus had shifted back to ARPA's original interest in missile defense. Years before, Ronald Reagan proposed a massive investment in missile defense; the agency, now renamed DARPA for Defense Advanced Research Projects Agency, had turned its interest to putting laser battle stations into space. Now that the United States had landed on the moon and NASA was building the space shuttle, perhaps the laser could become the ultimate weapon in space.

CHAPTER 4

SPACE LASERS ON THE HIGH FRONTIER

The end of the Apollo moon landings disappointed a generation of young space enthusiasts, but the years that followed brought a renewal of interest in space. NASA planned and began building the space shuttle to get the United States back into space. Visionary physicist Gerard K. O'Neill proposed human colonization of space. The Pentagon continued to expand their use of space for communications and reconnaissance, and began work on the global positioning system in 1973.[1]

In the early 1960s, ARPA had thought orbiting laser battle stations might be the ultimate weapon to counter nuclear missiles, but by 1965 it was painfully clear then available lasers were not up to the task. Interest in laser weapons turned to the battlefield. The 1967 Outer Space Treaty banned weapons of mass destruction from space but allowed other military uses, including laser weapons that directed their energy at specific targets, leading to growing concern about militarization of space.[2] However, military interest in battlefield lasers faded in the mid-1970s, and ARPA, renamed the Defense Advanced Research Projects Agency (DARPA) in 1976, was once again looking to make lasers the ultimate weapon in space.

Today, more than forty years later, that shift in direction might seem more like the actions of a hyperactive teenager than an agency charged with the future of military technology. If lasers were too massive and bulky for the battlefield, and struggled to shoot down nearby targets, why should anyone think they might work better in space? It's far harder to put a massive laser into orbit than to haul it to the battlefield, and targets in space are farther away than on the ground, at sea, or in the air.

However, times were different forty years ago. The United States and Soviet Union were armed to the teeth with nuclear arsenals in the dangerous standoff of the Cold War. During the heady years after men first walked on the moon, many people wanted to believe that we really were on the threshold of reaching the final frontier of space. And getting lasers above the atmosphere would avoid the tough problems of focusing laser beams through the turbulent atmosphere.

THE ARMS RACE AND THE SPACE RACE

The technological arms race of World War II that led to the atomic bomb continued into the Cold War years, where it gave rise to the space race and left the world bristling with nuclear arsenals with more than enough firepower to wipe out civilization. The quest was to find a new ultimate weapon that could defend against nuclear attack.

Airplanes dropped the first atomic bombs on Japan, but bombers take hours to reach their destinations and can be shot down by fighter jets or surface-to-air missiles. The United States and Soviet Union drew on the skills of their captured German rocket scientists to develop intercontinental ballistic missiles (ICBMs) that could carry nuclear warheads thousands of miles in a mere half hour. The United States put a crack team of rocket scientists under General Bernard Schriever to work on building intercontinental missiles in 1954, but they were far from finished at the time of the 1957 Sputnik launch.[3] What alarmed US officials about Sputnik was that a rocket powerful enough to put a satellite into orbit also could deliver a nuclear bomb halfway around the world, and that success seemed to indicate that the Soviets were ahead.

ARPA's failure to find a viable death ray to defend against nuclear attack led both sides to build massive nuclear arsenals. Both the United States and Soviet Union took a more conventional approach to nuclear defense, building fast interceptor rockets intended to destroy incoming ballistic missiles. The United States designed its Safeguard system to protect nuclear missile sites in Montana and North Dakota, so they could retaliate against a nuclear attack. The US system also included massive radar for detecting incoming missiles and other antimissile missiles tipped with five-megaton nuclear bombs that would detonate in space to destroy warheads before they entered the atmosphere. The Soviet Union built a system to defend Moscow.[4]

In 1972, the United States and Soviet Union signed the Anti-Ballistic Missile Treaty, which limited each country to one site within 150 kilometers of its capital and one within 150 kilometers of a missile site.[5] The US Congress eventually decided its system was so limited in capability that it was not worth finishing and shut it down in 1976.[6] The treaty allowed missile defense research to continue but banned the deployment of new antimissile missiles beyond the treaty limits.

The result was a deliberate stalemate called mutual assured destruction, or MAD, because both nations had thousands of independently targetable nuclear weapons on missiles on land and at sea, plus fleets of nuclear bombers, but only limited missile defense.[7] If one side launched an attack, the other would be devastated, but because defenses were limited, the victim of the attack would still be able to devastate the attacker in retaliation. Peace depended on knowledge, fear, and restraint.

The knowledge came from fleets of spy satellites that monitored launch areas and other sensitive sites in other countries. Spy satellite photos gave an essential ground truth of the other side's military capabilities. The fear came from the knowledge that no first strike could wipe out the other side's nukes, particularly those carried by nuclear submarines hidden and undetectable undersea. The result was a balance of terror that restrained both sides from pushing too far, although neither really trusted the other.[8] Arms control treaties were written to stabilize this uneasy balance.

With the stakes this high and the nuclear arsenals continuing to grow, it was no wonder that ARPA kept looking at prospects for a space-based defense against nuclear missiles in the 1970s. Its Soviet counterparts did the same thing. Laser technology had come a long way since the mid-1960s, including important new variations on rocket-engine lasers. By the mid-1970s, ARPA and the navy had developed the chemically fueled Navy Advanced Chemical Laser that could generate beams up to four hundred kilowatts, nearly halfway to a megawatt. That system ran on deuterium fluoride emitting at 3.8 micrometers, but in space a chemical laser could run on hydrogen fluoride emitting at 2.7 micrometers, which can't pass through the air.[9] That would shrink the focusing optics needed by almost a factor of four compared to the clumsy gas-dynamic rocket-engine laser running on carbon dioxide, another advantage in space. And the 1970s had brought a broad new optimism about the "high frontier" of space.

THE SPACE OPTIMISTS

The Apollo program's six successful moon landings were a remarkable triumph for the US space program. Less than a dozen years after the country had listened slack-jawed to the "beeps" from the Russian Sputnik I, the first artificial satellite to orbit Earth, two Americans had landed on the moon. The six successful landings spanned forty-two months, from July 1969 to December 1972. Space travel was no longer science fiction. It was a reality, promising real-world benefits from monitoring military activity around the globe to telecommunications and better weather monitoring and forecasting.

Rocket scientist Wernher von Braun had been the leading evangelist of space travel from the 1950s through the 1960s. Most of his public statements had focused on the exploration of space, but from World War II well into the 1950s, he also envisioned a military component, including space stations laden with nuclear weapons.[10]

A new group of visionaries had a different view of space, as a world to explore, colonize, and industrialize. Dreamers saw space as a new frontier, "the final frontier" that the TV show *Star Trek* explored in fiction. The most prominent was Gerard K. O'Neill, a Princeton University physicist who in 1969 started exploring the possibility of colonizing outer space itself rather than other planets in the solar system. He began calculating the possibilities as part of an exercise for ambitious students in his introductory physics class at Princeton. "As sometimes happens in the hard sciences, what began as a joke had to be taken more seriously when the numbers began to come out right," he wrote in *Physics Today* in 1974.[11]

O'Neill's calculations convinced him that "colonization is possible even with existing technology, if done in the most efficient ways. New methods are needed, but none goes beyond the range of present-day knowledge. The challenge is to bring the goal of space colonization into economic feasibility now, and the key is to treat the region beyond Earth not as a void but as a culture medium, rich in matter and energy."[12] He insisted that space colonization was possible "without robbing or harming anyone and without polluting anything." He envisioned "nearly all our industrial activity could be moved away from Earth's fragile biosphere within less than a century" from 1974. He expected that moving people and industry into space would "encourage self-sufficiency, small-scale governmental units, cultural diversity, and a high degree of independence." And he said the new space frontier could give humankind enough room to expand its population at least twenty thousand times above the 1974 level of four billion.[13]

His vision was enticing and helped stimulate formation of the National Space Institute in 1974. The L5 Society came the next year, getting its name from the Lagrange 5 orbital location in the Earth-moon system where O'Neill had proposed putting a colony.[14] The two later merged in 1987 to form the National Space Society, which continues to promote space exploration, development, and settlement, although those goals seemed much closer in the mid-1970s.[15]

O'Neill's optimism was far from unique. NASA's moon shot was a transformative experience for many people who lived through it. In 1957, outer space had been out of reach. By 1972, a dozen people had walked on the moon and lived to tell the tale. Those fifteen years had seen jets replace prop planes for air travel, transistors replace vacuum tubes, interstate highways spread across the United States, and computers become a big business. The pace of change was accelerating, Alvin Toffler wrote in his 1970 best-seller *Future Shock*.[16]

Planning for a fleet of reusable spacecraft that became the space shuttle had begun in 1969. The goal was to shuttle astronauts and equipment from Earth to orbit and back routinely. When science fiction writer Arthur C. Clarke and director Stanley Kubrick portrayed a routine trip to the moon on a Pan Am spaceplane in their movie *2001: A Space Odyssey*, it seemed more a projection of the future than science fiction. Plans in 1981 called for each of the four shuttles to fly at least once a month, a total of about fifty flights a year, which NASA never came close to achieving.[17]

New ideas emerged. Peter E. Glaser, a vice president of the consulting firm Arthur D. Little Inc., proposed building solar-power satellites to overcome energy shortages and price surges like those that had hit the United States in 1973 during the OPEC oil embargo. He envisioned giant satellites covered with solar cells that could beam gigawatts of clean energy down to Earth via microwaves or laser beams to receivers in deserts or other uninhabited zones (see photo insert). All, or at least most, of the mess and pollution of generating power would remain in space.[18] O'Neill proposed building giant electromagnetic mass drivers to catapult components of those giant solar satellites into geosynchronous orbit.[19]

ARPA SEEKS SHORT-WAVELENGTH SPACE LASERS

That bright outlook for space development led ARPA to reconsider space-based laser weapons in the 1970s. Ed Gerry arrived in 1971 and soon enlisted Peter Clark, a Caltech graduate who had been working at Hughes, to help launch new laser programs.[20] The first projects were small and focused on advancing laser technology to a point where it would justify support from other military agencies.

One goal was developing high-energy lasers with output at shorter wavelengths than the infrared beams of rocket-engine lasers. One important advantage came from a fundamental principle of optics. The shorter the wavelength, the smaller the spot a laser beam could be focused on; and the smaller the spot, the higher the light intensity. High intensity is important because the damage a laser beam can do to a target increases with the intensity. That means you need a bigger mirror to focus longer wavelengths to a high intensity; double the wavelength and you need a mirror twice as wide with four times the area, and big mirrors can be cumbersome. In addition, photons, the basic units of energy in light, have more energy at shorter wavelengths, and many materials absorb a larger fraction of the incident light at shorter wavelengths. People learn that lesson firsthand when they get a sunburn because the damage comes from short wavelengths in the ultraviolet range rather than the visible light we see.

Rocket-engine lasers emit light that comes from the vibrations of hot molecules like carbon dioxide and hydrogen fluoride. Those vibrations correspond to infrared wavelengths, and the atmosphere absorbs strongly in much of the infrared range because it's made of gas molecules that can absorb infrared light to make their own atoms vibrate. Carbon dioxide in the air absorbs the light from carbon dioxide rocket-engine lasers. The air contains virtually no hydrogen fluoride, but other molecules also absorb light in that band. One of the few infrared bands with weak absorption is around 3.8 micrometers where deuterium fluoride lasers emit but few molecules in the air absorb. The air is clearest at visible wavelengths, which is why animals evolved eyes that see in that band.

Weapons were not the only goal of ARPA's hunt for new lasers. The Pentagon needed a better way to keep in touch with its ballistic missile submarines when they were underwater. Only extremely long radio waves could penetrate the ocean, and they couldn't carry much information. However, ocean water does transmit blue-green light well, and ARPA sought lasers transmitting in that band to keep in contact with its submarines.[21]

BIG ZAP LASERS

In addition to inventing the rocket-engine laser, Avco physicists found they could produce powerful bursts of laser light from carbon dioxide by zapping the gas with a powerful pulse of electrons. Think of it as a better-controlled version of the lightning bolts that were the death rays of the ancient gods or the sparks that Nikola Tesla generated with the Tesla coil. Avco's Thumper and Humdinger lasers produced powerful beams of infrared light by firing blasts of electrons through carbon dioxide.

In 1971, Soviet physicist Nikolai Basov zapped the rare gas xenon with powerful electron pulses that ripped electrons from the outer shell of the atoms. Xenon normally is chemically inert because its outer shell of electrons is full, but blasting off some of the outer electrons let pairs of xenon atoms combine to form a short-lived two-atom molecule that emitted bright ultraviolet light at 176 nanometers.[22] That wavelength can't go far in air, which is a good thing because it damages DNA. But it showed a new approach to making a high-energy ultraviolet laser, so ARPA started investigating.

A breakthrough came in 1974 when researchers started zapping mixtures of rare gases with the highly reactive halogens fluorine, chlorine, bromine, and iodine. This produced short-lived molecules containing one atom of a halogen and one atom of the rare earths argon, krypton, or xenon. The bolt of electricity gave the two atoms enough energy to react to form a molecule that is born in an excited state but falls apart when it releases the extra energy in the form of ultraviolet light. A whole family of those lasers, called "excimers" for excited dimers, emerged in 1975, as mentioned in chapter 3.[23]

Excimer lasers could only generate pulses, but they were by far the brightest ultraviolet laser to date, so the agency began investigating them as possible weapon lasers and for shifting their output to blue-green wavelengths for communications from satellites or aircraft to submarines.

X-RAY AND FREE-ELECTRON LASERS

ARPA also explored prospects for two other novel lasers that seemed to have potential for high energy.

The next logical step beyond the short wavelengths of ultraviolet lasers was the quest for lasers emitting at even shorter wavelengths in the X-ray band. The shortest ultraviolet wavelengths reached in the early 1970s were around 110 nanometers.[24] X-rays are emitted on more energetic transitions that occur when electrons drop to energy levels deeper down in the inner shells of atoms. The X-ray band is often said to span wavelengths from ten nanometers to a tenth of a nanometer, a range from roughly one-fiftieth to one five-thousandth of visible light.

By the early 1970s, X-ray lasers were looking difficult, but that didn't stop experimenters from seeking fame, fortune, or at least a PhD by making the first one. In July 1972, University of Utah graduate student John G. Kepros claimed success with an oddly simple experiment. He had dissolved copper sulfate into grocery-store unflavored gelatin, spread the mixture onto microscope slides, and zapped the slides with powerful pulses from a glass laser lasting twenty billionths of a second. Standard packs of Polaroid film shielded by dark paper and metal foil showed dark spots exactly under the laser spots. Kepros claimed only light from an X-ray laser could have penetrated the wrapping.[25]

The report yielded a few headlines about a Jell-O laser breakthrough but received much scrutiny from skeptical physicists who could not see how the glass laser pulse could have delivered enough energy to make X-rays. After Kepros described his experiment at the University of Michigan some months later, grad student Irving Bigio grilled him about details. Then Bigio taped a similar film pack to the wall and tried to duplicate the experiment without a laser. "I walked across the room, yelled 'bang' (pretending to be a high-power laser) then walked back across the room (slightly dragging my feet on the carpet), picked up the film pack, and developed it. Lo and behold, there were a couple of spots on the film," he wrote in an email. Bigio thought it was static electricity that he had picked up from the carpet, as Kepros may have done during a cold, dry Utah winter, but Kepros wasn't convinced.[26] Most researchers agreed it was neither X-rays nor a laser.[27] Later Bigio and others noted that Kepros worked in the same chemistry department where two electrochemists announced their discovery of cold fusion in 1989.

ARPA invested some money in X-ray laser research but, like many of its programs, the project lasted only a few years and yielded few results. It was cut to free money for research on another new idea that seemed more promising, the free-electron laser in 1976, the year the agency was renamed DARPA.[28]

John M. J. Madey had come up with the idea in a laser physics course at Caltech in the mid-1960s. If atoms or molecules could emit light when they changed energy levels, he wondered, why couldn't electrons do the same even if they weren't attached to atoms? The professor teaching the course thought it might be possible and gave him a reference. The idea stayed with Madey after he moved to Stanford, and he eventually worked it into his dissertation. He envisioned extracting laser energy from a beam of electrons passing through an array of magnets that bent the beam back and forth in what he called a "free-electron laser" because the electrons were not attached to any atoms.[29] He spent four years working on an experiment before successfully demonstrating one in 1976.[30]

Figure 4.1. John M. J. Madey (*left*) working with Luis Elias on the first free-electron laser. (Photo courtesy of Chuck Painter / Stanford News Service.)

Some laser physicists had doubted Madey's ideas, but that demonstration tipped the scales at DARPA as well as among laser researchers. The free-electron laser also had two big attractions. It drew power from a beam of relativistic electrons, which physicists knew how to make and which had the potential for achieving very high power. Its laser wavelength depended on the spacing of the magnets and the energy of the electrons, so it could be tuned across a range from microwaves to X-rays by varying those parameters. That made it a better bet for DARPA than the X-ray laser, which had remained elusive.

THE SPACE LASER TRIAD

The agency had begun shifting to bigger programs, ambitions, and budgets after George H. Heilmeier took over as director in 1975. He expanded the scale of many programs, including high-energy laser weapons.[31] His new thrust would bring a huge increase in DARPA's budget, which had been around $200 million in the early 1970s to a peak over $800 million in 1984.

Heilmeier wanted DAPRA to show that high-energy lasers could meet weapon-system requirements. That was a much more ambitious goal than building a whopping big laser that could fire hundreds of kilowatts for a few seconds at a time on the ground. Only the Airborne Laser Laboratory had tried that before, and the air force was struggling to squeeze a big laser, pointing and tracking optics, controls, and a power system into the plane, let alone tracking and destroying targets once it got off the ground.

DARPA had similar ambitions for the triad of space laser programs they launched but initially had a different target. Satellites had become vital in military activity, from communications to monitoring the activity of current and potential enemies. Spy satellites helped maintain the delicate balance of power by observing the other side's activities and military systems. The United States wanted a system that could both defend US satellites and disable enemy spacecraft if necessary. The original plan called for putting half a dozen laser battle stations in low earth orbit that could both take out low-altitude Soviet satellites and defend US satellites. To evaluate those goals, DARPA envisioned building three crucial parts of a space-based laser weapon: a giant rocket-engine chemical laser, a large mirror to aim the beam, and a sophisticated tracking and pointing system to identify targets and deliver lethal energy. Initial tests were to be on the ground.

In mid-1980, after a first round of contractor studies for DARPA's space-based laser program, Secretary of Defense Harold Brown reordered military priorities, boosting the role of lasers and shifting the main focus to developing high-energy lasers for use in space.[32] Experiments with big lasers including MIRACL and the Airborne Laser Laboratory had made the problems of sending laser beams through the atmosphere painfully clear.

An effect called thermal blooming seemed inevitable when directing high-energy beams through the air. Even if the air looked clear, it would absorb a small fraction of the light, and that energy would heat the air mostly at the center of the beam. As air is heated, it expands to become less dense, and the less dense the air, the lower its refractive index. That makes the path of light passing through the heated zone bend outward, spreading out the laser beam and reducing its intensity, just the opposite of what you want in a laser weapon. To measure the effect, Pratt & Whitney built two miles of railroad track in the Everglades behind its Florida plant and ran a small train along the tracks as researchers swept the beam from a rocket-engine laser across the area.[33]

In addition, the subtle turbulence of winds and rising and falling air currents could bend the laser beam back and forth, an effect evident when you see heat rising above a hot parking lot or across the hood of a dark car on a sunny day. Research labs were working on ways to make optics that could change the shape of their surface over time to correct for turbulence and other effects, but that technology was still in early development.

Brown's long-term goal was to develop orbiting laser battle stations to protect US satellites and perhaps to provide a future defense against long-range nuclear missiles. He spoke of a time frame of seven to ten years.[34] Adding the task of defending against nuclear missiles significantly expanded DARPA's original goal of using space lasers to defend US satellites and threaten those of the Soviet Union. Missiles are considerably more difficult to destroy or disable with lasers than are satellites.

At the time, DARPA's directed-energy office also defined the three key elements of what it called the Space Laser Triad. DARPA officials stressed that their goal was to demonstrate the technology needed to build a weapon system, not to deliver an operational system. They started by asking two contractors to develop designs for each of the three elements, with plans to pick the better design for each element of the Triad.[35]

With their megawatt-class rocket-engine laser intended to be used in space, DARPA didn't have to worry about atmospheric effects, so they could use ordinary hydrogen fluoride, emitting at 2.7 micrometers, allowing optics only about a quarter of the size needed for a carbon dioxide laser emitting at ten micrometers. DARPA code-named the laser ALPHA. In 1980, they envisioned two stages of development, starting with a two-megawatt ground demonstration, with the eventual goal of a five-megawatt version of ALPHA compact enough to be used in space. The laser and other components were to be designed to fit into the space shuttle bay for launches in the 1990s.[36]

LARGE OPTICS FOR LARGE LASERS

A second key element of the Triad was demonstrating the feasibility of a space mirror big enough to focus a powerful laser beam onto a distant target with a high enough brightness to do lethal damage. A giant mirror was a must to concentrate the five-megawatt power from a space-based rocket-engine laser onto a small spot vulnerable to damage. In the late 1970s, DARPA decided that the best mirror size for this laser would be four meters. Nothing that big had ever gotten off the ground before, which made the Large Optics Demonstration Experiment (LODE) a DARPA-hard product.

The four-meter size of the mirror, like many choices in a military system, depended on the technology already developed and tested. The Pentagon had agreed to use the space shuttle for future launches, so everything had to fit into the shuttle's cargo bay, and four meters was as big a mirror as it could fit. But new technology was needed both to reach the four-meter size and to develop big mirrors able to withstand the megawatt-class beams sought for laser weapons without warping or melting.[37]

One option was to scale up the technology of the Hubble Space Telescope. At the time, the public saw Hubble as the cutting edge of large space optics, but in reality it was a tried-and-true technology developed for heavily classified spy satellites that the United States orbited to monitor Soviet military activity. However, the big optics companies didn't want to scale up their mirrors from the 2.4-meter standard for spy satellites. NASA had wanted a three-meter mirror, but when the space agency ran into a budget crunch, they offered a much better deal for the standard 2.4-meter size. They also resisted developing a four-meter version of the Hubble design for DARPA, recalls Doris Hamill, who evaluated the options for developing such a mirror as a young air force lieutenant at the Rome Air Development Center in upstate New York.[38]

A second option was to use a technology DARPA had studied for use in infrared spy satellites planned to be stationed in geosynchronous orbit, called High Altitude Large Optics (HALO). That required very large mirrors to see details on the ground, so DARPA envisioned using a single, thin glass sheet with little actuators in the back to keep the shape just right for imaging as satellites experienced changing conditions that might shift the mirror's shape. That particular system was never built, but it gave people ideas.

A third option was to make a segmented mirror, with a central hexagonal piece surrounded by six other segments. That would require aligning the segments so the overall surface was accurate to within a quarter of the wavelength of visible light, a few tenths of a thousandth of a millimeter. That sounded demanding. But Hamill's boss, Colonel Ronald Prater, finally settled on the segmented mirror because he believed it would scale more easily than either other alternative, and that US technology was well equipped to cope with the challenges of building the required control systems. That became the design for LODE. DARPA was impressed enough by Hamill's work to hire her as a program manager soon afterward.[39]

Each element was intended to push the limits of the available technology, but DARPA thought the biggest challenge would be Talon Gold, a system for tracking moving targets and keeping the laser beam pointing precisely at them. That system had to combine fast response to target motion with pinpoint precision in aiming the beam and maintaining the orientation of the laser.[40] A reflection of the importance and difficulty of Talon Gold was that it had been the only part of the Triad planned for on-orbit tests, with the first round planned for the space shuttle in 1984.[41]

The ideas were far from all new. The combination of a five-megawatt laser and a four-meter mirror had been proposed earlier in the 1970s.[42] Nor were the designs settled. DARPA continued looking at new types of lasers, including an innovative rocket-engine laser that worked with a different chemical brew in which chlorine gas reacts with a mixture of hydrogen peroxide and potassium hydroxide to form excited oxygen molecules that transfer their energy to iodine.[43]

DARPA's plan was for a series of systematic ground-based experiments to test the potential of the new technologies needed for the project. It would take years to perform all the tests, and it was expected to cost somewhere around a billion dollars.[44] It was already under challenge by groups far more optimistic about lasers and other space-based defense systems.

MAX HUNTER AND THE GANG OF FOUR

Outside of DARPA and the military bureaucracy, laser advocates had grown impatient with the slow progress of a technology they saw as tremendously promising. On Halloween day of 1977, Maxwell W. Hunter II put the finishing touches on a proposal titled "Strategic Dynamics and Space-Laser Weaponry," which would become a manifesto for laser advocates. It proposed a big push to deploy orbital laser battle stations to block a Soviet nuclear attack.[45]

Known as "Max" to friends and colleagues in fields ranging from aerospace to science fiction, Hunter was a veteran engineer who had earned a master's in aeronautical engineering from MIT. He started designing aeronautics and missile systems for Douglas Aircraft in 1944, moved to NASA in 1961, and moved to Lockheed Missiles and Space in 1965, where he would stay until he retired in 1987.[46]

He first encountered lasers in 1966 while heading a study of a plan called BAMBI (Ballistic Missile Boost Intercept) that proposed stationing interceptors on satellites to destroy nuclear missiles. His study group decided BAMBI was impractical because it would require hauling too much tonnage to orbit to be practical with current technology. They thought laser battle stations would be more affordable but didn't know about the supersecret rocket-engine laser and concluded that the needed lasers were yet to be developed.[47]

Hunter changed his mind once he learned about rocket-engine lasers and urged using them for space-based missile defense. The Pentagon supported paper studies of the prospects at Lockheed in the 1970s, and by 1977 Hunter had convinced himself that lasers could be an effective defense against strategic nuclear missiles, making them the ultimate weapon to end the nuclear stalemate. It was time to do something.

"I suddenly realized that lasers are something we hadn't tried before," Hunter said later. "It may be decades before we understand the full implications of a speed of light interceptor, but there's one thing you know: the best interceptors are the fastest; and until Einstein is proven wrong, lasers are going to be the fastest interceptors. So if you build up to where they have enough pizzazz to hurt something, you better back off and seriously consider where these weapons will take you."[48]

Like many engineers, Hunter was impatient with strategy, psychology, and diplomacy. He would rather build defense systems to block an attack than rely on the threat of massive retaliation to deter one. In his Halloween paper, he wrote that "high energy lasers are proliferating and space transportation is about to become sufficiently economical that, if it is used to place such lasers in space, an effective defense against even massive ballistic missile exchanges . . . is indeed possible." He was dazzled by the potential power of a speed-of-light defense, noting that light moves fifty thousand times faster than a rocket interceptor, making it a revolutionary and far more potent class of weapon. Lasers also had the advantage of firing tightly focused beams with pinpoint accuracy, making the risks of extensive damage from a mistaken laser shot far below the risk of a misdirected nuclear missile. That, he argued, would allow computers to run laser battle stations without human intervention, giving the laser weapons the lightning-quick reaction time essential for their mission.[49]

What made Hunter think of a space-based laser battle station as an ultimate weapon was its potential to zap nuclear missiles as they were boosting through the atmosphere, when they are most vulnerable to attack. The boost phase makes a tempting and relatively easy target to spot because it is hot (thanks to friction from the air and heat from the rocket engine) and thus easy for infrared sensors to spot from space. The booster also is a larger and somewhat more vulnerable target than a reentry vehicle hardened to withstand the heating from its reentry into the atmosphere. Destroying a booster knocks out all the warheads it carries, avoiding the need to destroy multiple small, cold, and hard-to-spot warheads deployed from the booster.[50]

Figure 4.2. Max Hunter, who envisioned a fleet of orbiting rocket-engine laser battle stations that could zap the entire Soviet nuclear missile arsenal. (Photo © 2018 Maxwell W. Hunter Foundation.)

Making such a laser battle station work would have required tremendous advances in computing and communications. A modern smartphone would have run rings around the world's fastest supercomputer in 1977. US satellites monitored continuously for the infrared signatures of Soviet missile launches in the 1970s, but data transmission rates were limited and the data had to be analyzed on the ground.[51]

All in all, Hunter's plan for robotic laser battle stations was a visionary scheme, assuming tremendous advances in lasers, computers, and communications. It came at a time of technological optimism spanning fields from lasers and computing to space travel. Hunter extrapolated optimistically from the technological trends of his time. His ideas were the sort that a science fiction writer might use as the background for a novel set in the near future, a decade or more away, but not centuries in the distant future. It's the sort of thing military technologists must do in planning, and it can make for a fascinating novel. But there is no way to be sure that it will predict the future.

SENATOR WALLOP AND THE GANG OF FOUR

Hunter circulated his paper among friends and colleagues. He also became an anonymous source for journalist Clarence A. Robinson Jr. who wrote for the influential industry magazine *Aviation Week & Space Technology*.[52] Hunter wanted to find someone who could make something happen beyond just another DARPA research project. He eventually found that person in the form of a Republican senator named Malcolm Wallop.

Wallop was a staunchly conservative Wyoming rancher and small businessman with an unusual background. His paternal grandfather, Oliver Henry Wallop, was a British aristocrat, whose older brother John was the 7th Earl of Portsmouth. Oliver Henry had immigrated to America and settled in Wyoming, where he became a prosperous cattle rancher and a member of the Wyoming House of Representatives. But when his older brother died without issue, Oliver Henry went back home and became a member of the House of Lords until his death in 1943.[53] His son Oliver Malcolm Wallop remained in America, where he married into an East Coast society family. The future senator, Malcolm Wallop, was born in New York City in 1933, then settled in Wyoming after attending Yale and serving in the military. After getting involved in politics, he was elected to the Senate in 1976.[54]

As a freshman Republican senator in a Senate controlled by Democrats, Wallop had little power. But that did not stop him from expressing strong opinions on strategic defense. Named to the Senate Intelligence Committee in 1977, he hired Italian-born Angelo Codevilla, a young political scientist who shared Wallop's concerns about strategic defense, to work on the intelligence committee staff.[55]

Their Senate intelligence posts made Wallop and Codevilla keenly aware of the uneasy balance of terror of mutual assured destruction. A new generation of electronic spy satellites was revealing a Soviet military buildup that alarmed them. They feared the United States lacked the antimissile defense needed to stop a Soviet first strike intended to knock out the capacity to respond with a devastating counterstrike, and were looking for others who shared their views. In the summer of 1978, Codevilla heard Hunter speak at a Washington conference on strategic defense and was excited by Hunter's ideas. His bold vision of space-based laser battle stations able to blow the whole Soviet nuclear arsenal out of the sky excited Wallop as well as Codevilla.[56]

Wallop took advantage of his Senate post to visit the army's missile defense group in Huntsville, Alabama, and ask probing questions. He found both new technology relevant to the task and officers growing impatient with the pace of progress. Some officers complained that concerns about violating the ABM Treaty were impeding missile defense research and that some army and air force groups were opposing missile defense to preserve existing programs.[57] He continued gathering evidence in 1979 and published a proposal in the fall 1979 issue of *Strategic Review* to build a fleet of eighteen laser battle station satellites. Six battle stations would circle the globe in each of three polar orbits, allowing them to cover the globe. Each would carry a five-megawatt rocket-engine laser, a four-meter focusing mirror, and battle management controls. Each laser would be able to fire a thousand shots with a range "somewhat less than 3000 miles." Wallop claimed those components "could protect against all Soviet heavy missiles, about 300 other ICBMS, nearly all [submarine-launched missiles], and all long-range bombers and cruise-missile carriers."[58]

The technology of the plan was largely Hunter's. But Wallop offered one sobering realistic observation about the Cold War arms race: "There is no ultimate weapon, but this one holds great promise for a decade or two of real protection for Americans."[59]

Wallop testified at Senate hearings on missile defense, and his proposals got into the news, but that failed to stimulate any action. He tried to enlist aid from industry, but executives demurred, worried about upsetting Congress. Two key officials of DARPA's laser weapon program declined to discuss development timetables. In the end, under pressure from Codevilla, Hunter agreed to prepare a talk outlining the proposal, and company officials agreed to supply three other experts to talk about the plan, as long as they said it was their personal opinion, not that of their companies'.[60] The result was a group dubbed "the Lockheed gang of four" (an allusion to a Chinese political faction): Hunter; TRW chemical-laser expert Joseph Miller; Perkin-Elmer optics expert Norbert Schnog; and Gerald Ouellette, a pointing-and-tracking expert from Charles Stark Draper Laboratory.[61]

Figure 4.3. Malcolm Wallop, the Senate's biggest enthusiast for laser weapons in the 1970s and 1980s. (Photo courtesy of the US Senate.)

The group provided some more interesting details. The as-yet-unfinished space shuttle would carry components into orbit, where they would be assembled and fueled. The laser would weigh in at about 37,400 pounds (17,000 kilograms) and measure nineteen to twenty-seven feet (six to eight meters) long. One or two shuttle flights would be needed to carry the fuel for each battle station. They estimated the whole system would cost $10 billion.[62] A Pentagon review was not quite as optimistic, estimating twenty-five battle stations would be needed to intercept a thousand Soviet boosters, needing ten to twenty seconds to destroy or disable each one.[63]

The big difference between Wallop and Hunter on one side and DARPA on the other was in the timing and goals of the program. The Pentagon envisioned spending several years on research before building anything. Wallop and Hunter wanted a bigger-budget crash-priority program.[64] They also wanted to expand goals of the Triad program from DARPA's plan for defending and attacking satellites to Hunter's plan to target boost-phase ballistic missiles, much more difficult targets that would require higher laser intensities to destroy.

Wallop's effort to triple funding for the Triad program in the fiscal 1980 budget was defeated.[65] However, he did make a vital connection with Ronald Reagan late in 1979 when Reagan was laying the groundwork for his 1980 presidential campaign. Reagan shared Wallop's dislike of mutual assured destruction and believed that strategic defense including new technology to block nuclear attack would be a better option.[66]

CHANGE AFTER REAGAN'S ELECTION

Reagan's election brought change and a reevaluation of laser weapon programs. In early 1981, the Senate Armed Services Committee asked the Pentagon to analyze prospects for accelerating deployment of space-based laser weapons.[67] The Pentagon responded that "revolutionary" change might be possible but warned of large uncertainties about ultimate capabilities and costs.[68] The report prepared by the Pentagon called ballistic missile defense "the most demanding application" for space lasers. Key uncertainties included how well space-based lasers could withstand enemy attack, and the difficulty of killing future generations of ballistic missiles that could be designed to withstand laser attack by the time the laser battle stations were deployed in orbit. Not until 1994 could a space-based laser weapon be deployed, the Pentagon report said, and even then it would be "largely limited to antisatellite missions and would have no growth potential to satisfy the more demanding missions such as antiaircraft or ballistic missile defense." Projected cost of the initial demonstration would be in the $10 billion range.[69] Actually deploying a system of eight battle stations capable of shooting down planes would cost $25–$55 billion.[70] The study projected that ballistic missile defense would require 54–285 laser battle stations and cost $100–$800 billion, depending upon assumptions that were redacted in the declassified report.[71]

Wallop pushed for increasing the budget and speeding up the program, and under the first budget passed by the new Congress, DARPA's budget for laser weapons jumped by more than two-thirds, to $108 million. A smaller increase followed in the 1983 budget.[72]

Goals for the Space Laser Triad were expanded to include missile as well as satellite defense, but the hardware wasn't upgraded to keep pace. "The Triad got repurposed into being a missile killer, which it couldn't have done," said one veteran laser weaponeer. A five-megawatt beam focused by a four-meter mirror could take out satellites, but not the Soviet SS-18 "Satan" nuclear missile. Planners talked about eventually scaling up to ten-megawatt lasers with ten-meter mirrors or twenty-five-megawatt lasers with fifteen-meter mirrors. That would have required developing segmented mirrors that could be unfolded in space, but planners thought it would have been more capable of the antimissile mission. However, such giant space mirrors were beyond the Pentagon's budget.

GENERAL GRAHAM AND THE HIGH FRONTIER

Reagan's military advisor for his 1976 and 1980 presidential campaigns also was a strong advocate for strategic defense. Retired army lieutenant general Daniel O. Graham had a background in intelligence, including stints as deputy director of the Central Intelligence Agency and head of the Defense Intelligence Agency.[73] He also was enamored with space, which he saw as the ultimate high ground for missile defense. But he was not enamored with lasers or other high-tech directed-energy weapons. He wanted technology that was ready to deploy, and to him that meant interceptors that would kill nuclear missiles by ramming into them.

Instead of joining the Reagan administration, Graham founded a group called High Frontier and launched a study sponsored by the Heritage Foundation to develop a plan for missile defense. A central concept was a layered defense, in which attacking nuclear missiles would have to run a gauntlet of multiple defense systems to reach their targets. He championed defenses that could be built with existing technology, and made his top priority the quick deployment of a ground-based system that could fire a barrage of projectiles at incoming warheads that came within a mile of a missile silo.[74]

The next step in his plan was orbiting battle stations bristling with simple nonnuclear projectiles that could destroy Soviet missiles. Sometimes called "smart rocks," they were small projectiles powered by rocket engines with guidance systems that would steer them to ram into a missile or warhead, disabling or destroying it. Like Hunter and Wallop's space lasers, Graham's High Frontier would require a fleet of orbiting battle stations to continuously cover potential missile trajectories. He hoped to have it deployed in five to six years at a cost of $10–$15 billion. A second generation of space-based projectiles, or possibly lasers or other beam weapons, would follow.

To carry all that gear into space, he proposed building a high-performance spaceplane that could carry soldiers into space, retrieve or repair satellites, and inspect any spacecraft that worried the Pentagon. He also pushed other space projects, including upgrading the space shuttle, creating a cargo version, and developing a next-generation spacecraft with larger capacity. High Frontier also had a civilian side, which proposed building a space station, aiding industrial development in space, and building a space solar-power satellite.[75]

THE QUEST FOR NUCLEAR X-RAY LASERS

When DARPA gave up on X-ray lasers in 1976, two young physicists at the Lawrence Livermore National Laboratory were just getting started. George Chapline and Lowell Wood knew that the X-ray laser could only work under very unusual conditions, but they also realized that X-rays could be an extraordinarily sensitive probe of atoms, molecules, and crystalline structure. Rosalind Franklin had famously used the scattering of ordinary X-rays to deduce the double-helix structure of DNA, but the coherent X-rays from an X-ray laser could greatly increase the sensitivity of measurements and allow three-dimensional X-ray holography of biological molecules and DNA in cell nuclei, yielding much more information than Franklin had been able to collect, Chapline and Wood wrote in 1975.[76]

Other groups also were seeking X-ray lasers, but Livermore had two big advantages. One they described in *Physics Today* was the giant pulsed lasers that the nuclear laboratory was building for experiments in nuclear fusion. Focusing a laser pulse that for a billionth of a second delivered a trillion watts of power onto a tiny spot could rip inner shell electrons out of atoms, creating the right conditions for X-ray lasing. No X-ray mirrors existed to make a laser resonator, but the two thought they might be able to produce a thin laser-like beam if their laser pulses could form a long, thin region of highly ionized plasma.[77] The other advantage was in their minds then, but they didn't talk about it, because the details were classified; it was Livermore's role in designing and testing nuclear bombs.

The idea of making an X-ray laser came to the two while they were walking in the foothills above the lab. "Congress had asked the labs, isn't there something more interesting you can do with the nuclear weapons projects than just come up with another warhead design?" Chapline told me later. "During that walk, we thought that pumping an X-ray laser would be an interesting thing to talk about."[78]

Chapline, a theoretical physicist who studied under Nobel laureate Murray Gell-Mann and graduated from Caltech in 1967,[79] didn't know how to solve the problem then. Later, at a physics meeting in Novosibirsk, Russia, he heard Russian physicist I. I. Sobelman explain how laser action could be excited in plasmas. Later, listening to a description of how a nuclear weapon emitted powerful bursts of X-rays made him realize that those bomb-produced X-rays might produce the plasma he needed. "Hearing the results of the experiment, I instantaneously put together the ideas I had gotten from Sobelman's talk with the results of the experiment, and in five minutes came up with the general idea of something that would most likely work. It became clear I could make an X-ray laser with a nuclear device," he said.[80]

The X-rays come from matter heated to million-degree temperatures in a millionth of a second by the energy from the nuclear blast. At a million degrees, the nuclear fireball radiates X-rays like hot volcanic lava glows red with heat.[81] Chapline realized that intense bursts of X-rays could tear electrons out of atoms to produce a hot plasma that could produce X-ray lasing as the excited electrons dropped down to a lower energy level inside the atom. Soviet physicist O. N. Krokhin came up with the same idea independently at some point, but that work was done under security wraps.[82]

Wood, a bearded protégé of Edward Teller with a reputation for wild ideas, already was interested in laboratory-scale X-ray lasers. Yet, Wood initially was skeptical about prospects for a bomb-driven X-ray laser. Then Peter Hagelstein, a younger physicist from MIT who had been working on a laboratory X-ray laser at Livermore, came up with a different way to make a bomb-driven laser when reviewing Chapline's idea. They realized that a single experiment during a nuclear test could evaluate both approaches. "Then Lowell became a proponent, and he got Teller pumped up about the whole business," Chapline recalled.[83]

The bomb-driven X-ray laser was essentially classified at birth because it involved two highly classified fields, nuclear testing and some details of nuclear weapon design. Nuclear weapon theory, design, effects, and testing are interactive processes. Livermore and other laboratories have developed sophisticated computer codes that model the physics of nuclear explosions, and nuclear tests were conducted to validate the codes and gather new information. However, nuclear tests were difficult and expensive, and the tremendous energy released in a millionth of a second made measurements difficult and destroyed the test equipment in the underground chamber where the tests were conducted. That meant tests were few and far between, and the results didn't always work as had been hoped. The bomb-driven laser experiments came first, presumably because they were better prospects for use as weapons.

Chapline described the material used in the tests as "an organic pith material" from a weedy plant growing on a vacant lot in nearby Walnut Creek, California. Pith is a light, thin-walled cellular structure usually found at the center of stems. It was later replaced by aerogels, ultralight porous materials.[84]

The first test of the nuclear X-ray laser was an add-on to a 1978 test conducted by the Defense Nuclear Agency, which was unsuccessful because the vacuum system failed. Two years later a test code-named Dauphin, devoted entirely to the X-ray laser, yielded encouraging results. Both the Chapline and Hagelstein designs seemed to work, but Hagelstein's was chosen for further work because it looked more promising for missile defense.[85]

Others at Livermore kept working on laboratory-scale X-ray lasers, but Teller and Wood focused on the bomb-driven version. Even extremely powerful X-ray lasers would be useless as ground-based weapons because X-rays can't travel far in air. Thus Teller and Wood sketched out plans to zap ballistic missiles in space with their new bomb-powered death rays, either stationed in orbiting battle stations or housed in ground-based missiles that could pop up into space and detonate there.[86] Teller called it a third-generation nuclear weapon (the atomic bomb was the first generation, and the hydrogen bomb the second) because it promised a way to concentrate its lethal force in certain directions or types of energy, rather than explode indiscriminately to destroy everything in the area. The prospect of concentrating the tremendous energy from the nuclear explosion excited Teller, who may have seen it as a way to control the power he had unleashed in the hydrogen bomb.

The highly classified results of the test were soon leaked to *Aviation Week* reporter Clarence Robinson, whose three-page report led to widespread coverage elsewhere in February 1981.[87] Robinson's story claimed that the X-ray pulse of several hundred trillion watts lasted for around a billionth of a second. Its wavelength was measured at 1.4 nanometers, about three hundred times shorter than visible light. But he did not mention how rapidly the X-ray beam spread after emission, a crucial factor for laser weapons. A beam that spread broadly like a searchlight would have been of little use.

The big news for most observers was Livermore's plan for making X-ray laser weapons. Bomb-driven lasers would inherently be one-shot weapons, with only an expanding cloud of radioactive debris remaining after the explosion. One idea was surrounding a nuclear warhead with a ring of about fifty laser rods, pointing all the rods simultaneously at Soviet missiles, then detonating the bomb to produce a powerful X-ray pulse that would excite powerful X-ray pulses from each of the rods. Those pulses supposedly would carry enough energy to destroy the target by blasting material from the surface. Twenty to thirty X-ray laser battle stations supposedly would suffice to blow away a Soviet nuclear missile attack lasting thirty minutes. But a spread-out attack could deplete the X-ray laser arsenal, so other defenses would be needed.[88] The leaked story said the X-ray battle stations would be stationed in orbit permanently, which would have left them vulnerable to antisatellite missiles. Later Teller and Wood said the bomb-driven X-ray lasers could be kept on the ground and launched on warning of attack to blow away enemy missiles.

Livermore code-named the X-ray laser Project Excalibur, after the legendary sword of the equally legendary King Arthur of England. It was news to most of the military. Department of Energy officials were furious about the leak. Livermore ordered all its employees not to comment on the news story or the test. Pentagon officials similarly were told not to comment.[89] Just before the leak, Teller and Wood had come to Washington, where *Aviation Week* was based, to brief congressional leaders on the X-ray laser.[90] A third Livermore scientist, Roy Woodruff, had come along to keep Teller and Wood from overselling their plan.[91] The timing suggested the leak came from someone involved in the briefings, with Wood considered the most likely suspect, but no source was ever publicly identified.

Both the concept and the interpretation of the experiments would prove controversial. But the bomb-driven X-ray laser did join space-based rocket-engine lasers and High Frontier as concepts in serious discussion for use in strategic missile defense.

SUPERMARKET SCANNERS AND CD PLAYERS

The laser industry was small when I started working at *Laser Focus* magazine in 1974. It was a small magazine, with readership under twenty thousand, and industry sales for all of that year were estimated at $267 million. One hot topic was developing laser scanners to read bar codes on food packaging, although they did not become standard in supermarkets until the early 1980s. Another was laser medicine, where a laser treatment of the retina had become standard to stave off diabetes-related blindness, leading to sales of $5 million worth of lasers to perform the treatment.[92]

As the junior editor, I wrote brief descriptions of new products. Mostly they were meaningful only to readers and advertisers, but one that stuck in my mind from 1975 was the first commercial semiconductor laser that could emit light continuously at room temperature. Just five years earlier, getting such lasers to emit continuously for seconds had been a breakthrough; in 1975 they reliably lasted thousands of hours but still emitted only milliwatts. In 1976 the cutting edge of performance became one-hundred-year lifetimes extrapolated at Bell Labs. I never expected to see the semiconductor lasers power weapons.

In 1980, sales of lasers and related equipment passed a billion dollars a year, with an estimated $200 million going to military research, largely in the classified realm of laser weapons. Sales of high-powered civilian lasers for cutting, welding, drilling, and heat treating grew to $70 million, up 25 percent.[93]

Bill Shiner, who started as a technician with American Optical in 1962, mortgaged his home in 1973 to buy the company's laser division with Al Battista, the engineering manager. American Optical had been sold in 1967, and the new owners were giving up on lasers. But Shiner and Battista knew of two emerging uses of lasers for specialized machining processes. Sales took off when aircraft companies started using lasers to drill holes to improve air flow in jet engines.[94]

People grew with the laser industry. In 1974, Hughes Aircraft lured Hal Walker from Korad to manage development, design, and testing of military laser and electro-optical systems. Hughes was manufacturing laser range finders and other optical equipment for pinpointing targets, seeing at nighttime, and other missions. The lasers weren't powerful enough to kill anyone with their light, so they weren't death rays, but the laser beams marked enemy targets with codes that smart bombs or missiles could home in on.

Walker became the highest-ranking African American in Hughes management. In that position, the company asked him to share an office in the laser physics department with a young African American who had just earned a doctorate in laser physics from MIT, Ronald McNair. The laser world was small in the 1970s, and so I happened to know McNair, who had written an article for me at *Laser Focus*. He joined Hughes as an engineer, but his real interest was in research, and he soon moved to Hughes Research Laboratories in Malibu, where Ted Maiman had made the first laser.[95] NASA selected him for its astronaut program in 1978, and he flew his first mission in February 1984 on the space shuttle *Challenger*. He died just seconds into his second shuttle flight when *Challenger* broke up on January 28, 1986.

CHAPTER 5
THE "STAR WARS" WARS

I call upon the scientific community in our country, those who gave us nuclear weapons, to turn their great talents now to the cause of mankind and world peace, to give us the means of rendering these nuclear weapons impotent and obsolete.

Tonight, consistent with our obligations of the ABM treaty and recognizing the need for closer consultation with our allies, I'm taking an important first step. I am directing a comprehensive and intensive effort to define a long-term research and development program to begin to achieve our ultimate goal of eliminating the threat posed by strategic nuclear missiles. This could pave the way for arms control measures to eliminate the weapons themselves. We seek neither military superiority nor political advantage. Our only purpose—one all people share—is to search for ways to reduce the danger of nuclear war.

—Ronald Reagan, March 23, 1983[1]

Ronald Reagan's "Star Wars" speech aired on television while Ed Gerry was waiting at Wright Patterson Air Force Base in Dayton, Ohio, for the arrival of his friend and colleague Art Gunther, chief scientist of the Air Force Weapons Laboratory. It was an event he would remember decades later. For Gerry, who was the president of W. J. Schaefer Associates—a defense contractor that was comparing the weapon prospects of space-based lasers, particle beams, and high-performance interceptors—Reagan's speech marked a sea change that would become evident the next day. He and Gunther had a talk scheduled then on missile defense technology. "We expected a few people, but when we got there, people were hanging from the rafters," Gerry recalls.[2]

Two years and two months into his presidency, Reagan said nothing about lasers or death rays when he announced his change in the nation's nuclear defense strategy. In his early career as an actor, he had saved a death-ray superweapon from saboteurs in a 1940 movie titled *Murder in the Air*.[3] In real life, he wanted to save the world from the nuclear weapons that had proliferated to the point of threatening Armageddon since the bombings of Hiroshima and Nagasaki. He hoped new technology could do that by offering a way to stop nuclear missiles far from their targets.

It had been over a decade since the United States and Soviet Union had signed the Anti-Ballistic Missile (ABM) Treaty, agreeing to limit the deployment of missile defenses. Yet Soviet arsenals had kept on growing. Reagan hoped that developing new missile defenses might encourage a wind-down of the nuclear standoff. Many other conservatives didn't trust the Soviets and wanted to develop a strong missile defense system that might be the ultimate weapon to end the nuclear balance of terror called mutual assured destruction. After the 1980 elections shifted the balance of political power to the Right, the conservatives began looking for technologies that could offer viable alternatives. The three approaches described at the end of chapter 4 were in the lead.

The Soviet establishment, keenly aware that they were behind the United States in key technologies including computing and electronics, worried that American scientists had the resources and knowledge to develop a missile defense system. American scientists, on the other hand, were stunned and hadn't a clue how to overcome the formidable obstacles.[4]

COMPETING VISIONS FOR SPACE AS THE HIGH GROUND

Long before Archimedes, armies had learned that occupying the high ground could give them an important advantage in war. They could look down from the heights to survey the enemy's positions. Anything they threw or fired downhill would go farther than what the enemy might hurl uphill.

The air was the high ground for both World Wars, but Wernher von Braun's V2 rocket had shown the potential of looking beyond the air into space. Sputnik took the Soviets to the high ground of orbit first, but the United States had caught up and claimed the lead in 1969 with the Apollo moon landings. By then space had proved to be a valuable high ground for spy, communications, and weather satellites. Both civilian and military activity grew through the 1970s, and the space shuttle had made its flight soon after Reagan's term began.[5]

Three main groups had begun pushing missile defense projects as the Reagan administration was settling in.

General Graham's High Frontier program pushed space industrialization as well as militarization and was allied with the Heritage Foundation, which had strong ties to Reagan. Their plan to launch battle stations bristling with high-speed interceptors built on existing technology offered a head start. But the ideas were old and had not gotten off the ground before, and opponents argued that Soviets could disable and destroy the satellite fleet before it could become operational.[6]

Wallop and Hunter's scheme for chemical laser battle stations pushed new technology more aggressively. However, the Defense Science Board concluded in April 1981 that space-based chemical lasers were too immature to attempt the integrated demonstration of missile defense that Wallop and Hunter had pushed. Wallop also had offended the political pecking order by attempting an end run around the Senate Armed Services Committee.[7] Moreover, space-based laser battle stations were as vulnerable to Soviet attack as satellites laden with interceptors.

Teller and Wood had pitched bomb-driven X-ray lasers to Congress and the Pentagon in February 1981,[8] and the leaked story had put the idea into public play. Teller was well-known in Washington, where he had been the top scientific advocate for all things nuclear since the 1950s. He and Wood had a bold new idea. But the X-ray laser's reliance on a nuclear weapon was a big minus, and Teller brought a mixed reputation and many enemies.

THE TELLER FACTOR

Edward Teller was a deeply controversial physicist who played an influential role in launching Reagan's missile defense initiative. He was sometimes called the inspiration for the mad nuclear weapon scientist Dr. Strangelove in Stanley Kubrick's brilliantly dark 1964 film satirizing the Cold War. Peter Goodchild even titled his 2004 biography *Edward Teller: The Real Dr. Strangelove*.[9] But watch Peter Sellers play the mad scientist in the film and you see an ex-Nazi who mistakenly salutes the US president (also played by Sellers) as "Mein Führer." Teller was far more complex and multifaceted.

Born to Jewish parents in Budapest in 1908, Teller found his early life was disrupted by World War I, a subsequent Communist coup in Hungary, and finally by the rise of the Nazis in Germany. After training in theoretical physics, he moved to the United States in 1935 to teach physics at George Washington University.[10] In his early years in America, he made important contributions to theoretical physics. He joined the Manhattan Project to help the US war effort, where he worked with many other American and immigrant physicists, and later played a key role in developing the hydrogen bomb. His testimony against renewing the security clearance of J. Robert Oppenheimer in 1954 alienated his former colleagues and the rest of the physics establishment, and he spent most of the rest of his career on military-related nuclear physics.

That made Teller a polarizing figure. Harold Brown and Michael May, who, like Teller, headed the Livermore lab, described the effect well in an obituary:

> Many regarded him as an unthinking advocate of nuclear weapons in particular, and weapons systems in general, as the answer to all questions of national security. To others, he was a creative architect of US military strength, a perceptive analyst of the international scene, and an accurate anticipator of future threats.

Teller earned his reputation for uncritical nuclear advocacy by proposing schemes that today seem worthy of Dr. Strangelove. His Project Chariot in the late 1950s envisioned using "peaceful" nuclear explosions for giant excavation projects, starting with blasting out a new harbor on the Alaskan coast and going from there to digging a new Panama Canal and Suez Canal. He got government backing but was blocked by grassroots opposition from scientists, Native peoples, and other Alaskan residents.[11]

Teller was well-known and well-connected. He cofounded the Livermore lab and remained a considerable force there for decades without having a line management responsibility. He became by far the country's best-known nuclear physicist, with considerable clout in military policy and nuclear energy.

In his public persona, Teller was a big-picture theorist who usually didn't get down to details. Once he realized how much earth could be moved by a nuclear explosion and how much easier that would be than conventional digging, he charged ahead with Project Chariot, glossing over smaller questions like environmental impact and nuclear fallout that in the end doomed the plan.[12]

"Teller's ideas were always big ideas, and he tended to proclaim each new Livermore project as the solution to one of the world's great problems," wrote Frances Fitzgerald. "Quite naturally, not all of his ideas seemed entirely sensible to others, and, just as naturally, not all of the projects Livermore undertook panned out in the way he advertised them."[13] One example was the quest for laser-induced nuclear fusion for energy production, launched at Livermore in the 1960s as a spin-off of the modeling of thermonuclear explosions. Livermore initially estimated that one kilojoule of laser energy would suffice to ignite fusion,[14] but the giant National Ignition Facility failed to ignite a fusion plasma despite 1.8 million joules of energy.[15] Critics often complained that Teller would say ideas were ready for engineering when they were far from worked out.

Teller had a reputation for being difficult, but he also could be charming and had a charisma that could attract young physicists. I heard both sides of his character the one time I interviewed him on the phone for a news story in the mid-1980s. He returned my call from an airport and opened with a question: was I related to Marjorie Mazel Hecht?

I immediately recognized what he was after. Marjorie Mazel Hecht was an editor at *Fusion* magazine, published by the Fusion Energy Foundation, which was headed by perennial presidential candidate Lyndon Larouche. They were enthusiastic supporters of nuclear fusion and laser weapons, but they had a penchant for conspiracy theories and other strange ideas that had gotten them banned from the premises at Livermore. I said no, and for a few minutes we shared our encounters with the Fusion Energy Foundation and had a good chuckle about their doings. Teller went on to say they had some good slogans, like "Feed Jane Fonda to the whales," which appealed to him because he somehow blamed her for causing the heart attack he had suffered three days after the Three Mile Island reactor accident in 1979. The Left-leaning actress had starred in *The China Syndrome*, a disaster movie about a fictional nuclear reactor accident released just twelve days before the real accident at Three Mile Island, and she was a harsh critic of nuclear power. (In his autobiography, published many years later, Teller blamed the heart attack on overwork as he tried to figure out what caused the accident.[16]) Then we went on to the business at hand, which I have long since forgotten.

That conversation gave me a window into Teller's mind. He could be pleasant, share anecdotes, and talk about ideas that interested him with a genuine enthusiasm, but it helped the conversation to have some background in physics and technology. It certainly helped to share our odd encounters with the Fusion Energy Foundation, and that I wasn't trying to pin him down on any tough issues. But I could also sense a touch of arrogance and some impatience with people who disagreed with him.

Talking about Teller with people who worked with him or reading what they wrote about him tends to give a more polarized view. To some he was a father figure who gladly shared his time and wisdom in a thick Hungarian accent and was always concerned about the good of the country. To others he was a dark figure, arrogant and dismissive of others' opinions while promoting his own interests, and utterly blind to his own overoptimism and errors. It was as if Teller were some sort of quantum mechanical entity with an internal duality so that observing him over a long period of time would yield a polarized view of one side or the other of him.

Figure 5.1. Ronald Reagan and Edward Teller in January 1982, on the road to Star Wars. (Photo courtesy of US government.)

Teller's personality made him influential in what would become the wars over "Star Wars." But his real power was more limited than he liked to think, his years were beginning to catch up with him, and his protégé Lowell Wood took over much of the job of promoting the X-ray laser as a weapon, putting Wood in the spotlight and making him a controversial figure.

Wood lacks Teller's depth as a physicist, but he can be a font of ideas, and has a broad range of interests. I was surprised to meet him at a paleontology conference, where he explained how laser pulses that flash intensely bright for less than a trillionth of a second can remove hard rock that entombs soft fossils, and we stood among the fossil hunters, chatting like two laser geeks.

Wood was only a few years older than the cadre of young physicists who did most of the work on the X-ray laser, and he built a close camaraderie with them.[17] He worked tirelessly promoting the X-ray laser.[18] But his hard sell and partisanship had led to distortions, and his reputation had suffered. He is the only person I ever saw who was hissed at when giving a talk at a scientific meeting, the First International Conference on Near-Earth Asteroids in San Juan Capistrano, California, in 1991.[19]

THE SCIENCE FICTION CONNECTION

One of the places Wood's travels took him was to the initial meeting of the Citizens Advisory Council on National Space Policy established by the American Astronautical Society, a professional society for aerospace engineers and scientists, and the L5 Society, a space advocacy group. Some thirty top space advocates and experts met January 30 to February 1, 1981, both to formulate a short-term policy to keep the space program healthy in the first year of the Reagan administration and to develop a long-term policy for the coming two decades. Jerry Pournelle, a member of the L5 Society board and former space scientist better known as a science fiction writer, chaired many of the sessions held in the home of another science fiction writer, Larry Niven.

Although Pournelle and Niven are best known for writing fiction, they took space policy as seriously as the academics, astronauts, and aerospace engineers at the meeting. Their goal was to get humanity into space, and they were worried by Reagan's plans to cut $600 million from the NASA budget for the coming year.[20] A policy paper followed a few months later, laying out the group's position:

> Space is potentially our most valuable resource. A properly developed space program can go far toward restoring national pride while developing significant and possibly decisive military and economic advantages.[21]

The panel was well aware of laser weapon development and considered space lasers to be "the best-known" emerging Soviet threat in space. But the overall assessment of beam weapons shows the fingerprints of Wood, who attended the meeting.[22] It mentions rocket-engine lasers and claims that a single one in high earth orbit could potentially destroy all the missiles the other side launched during a nuclear war, leaving the country with its laser on the high ground of space to rule the world. But then it says that the other side could easily protect its missiles from rocket-engine laser attack, making the rocket-engine laser battle stations "veritable sitting ducks."

Instead, the report recommended a different ultimate weapon:

> Pulsed space laser[s] energized by nuclear weapons exploded nearby—lasers which have been demonstrated by the U.S. in underground tests and in whose development in the Soviet Union is widely believed to be several years ahead—may be effectively impossible to countermeasure. They deliver too much energy of too penetrating nature in too short a period of time to defend against by any means known at present.
>
> These defensive weapons are kept in hardened silos, to be launched as soon as an enemy ICBM attack is detected. Such nuclear weapon pumped laser systems could fire lethal bolts of energy at dozens to hundreds of enemy missiles and warheads simultaneously, but would not have to defend themselves from attack beforehand. A dozen such bomb-energized laser systems—each launched by a single booster—could shield their owner's home territory from enemy attack for the half-hour period needed for its owner's ICBMs to be launched at, fly to, and destroy the enemy's missile and bomber fields.[23]

Those words presumably came from Wood, the only expert on laser weapons then on the panel. Later, General Graham and Max Hunter joined the panel.[24]

The back-and-forth went on through 1982. At Teller's recommendation, Reagan named George Keyworth as his science advisor. The science establishment worried because Keyworth was a little-known physicist from the Los Alamos National Laboratory, but his experience with lasers and nuclear weapons would be needed to assess missile defense options. However, Keyworth was more of a realist than Teller, saying the vital task was zapping missiles in the boost phase. "That is a formidable task and the technology is not in hand today. I would claim that self-pronounced laser experts who claim that it is something that we are a few years away from doing are plain not supported by the scientific and engineering communities. It is nothing better than speculation, and I think it is unsound speculation."[25]

On the other hand, Teller lobbied intensely for a Manhattan-scale project and raged at Woodruff for telling a review panel that it would take six years to assess Excalibur's scientific feasibility before deciding to develop a weapon. Teller wanted a "fully weaponized" X-ray laser in five years.[26]

Keyworth tried to keep Teller away from Reagan, whose aides tried to keep him away from special pleading. But Teller pulled his own strings and managed to wrangle an appointment on September 14, 1982. Keyworth called the result a "disaster."[27]

THE STAR WARS SPEECH AFTERMATH

Reagan's speech said nothing specific about technology. What he wanted was new technology for missile defense, and conservatives were happy with that. Details were not Reagan's thing. His speech worried Democrats and the establishment, who feared that trying to build missile defense would dangerously upset the balance of power and heat up the arms race, pushing the Soviets to increase their nuclear arsenals.

Critics scored an important political point the day after the speech when Massachusetts senator Edward Kennedy called Reagan's missile defense "reckless Star Wars schemes."[28] The "Star Wars" label captured a central reality: all the proposed missile defense plans were more fiction than science, based on risky technology not yet scaled to the point needed. As a catchy link to a top science fiction franchise, it caught the popular imagination. Later, it was officially named the Strategic Defense Initiative (SDI), but only military officials and contractors trying to hustle Pentagon money used the politically correct SDI. The press and public just called it "Star Wars," to the annoyance of George Lucas.[29]

The ensuing controversy echoed through the Reagan administration. The most profound worry was whether developing missile defense might destabilize the global balance of power enough to trigger a nuclear war. Another worry was whether it was possible to build a leak-free global umbrella to block nuclear ballistic missiles and, if so, what that would imply. Many political advocates claimed it was possible; the engineers knew better.

LAUNCH OF THE STRATEGIC DEFENSE INITIATIVE

President Dwight Eisenhower established ARPA on February 7, 1958, just over four months after the Sputnik I launch, although the first director, Roy Johnson, began with a very minimal staff.[30] President Reagan needed more than a year after his speech to name air force lieutenant general James A. Abrahamson, director of the Space Shuttle Program, to head the new Strategic Defense Initiative Organization (SDIO) in April 1984. His NASA office staff sent him off to his new job with a toy lightsaber, which he apparently wielded with good humor.[31]

Abrahamson immediately had his hands full. His billion-dollar 1984 budget was supposed to double each of the next two years, to four billion dollars in 1986. Reagan wanted a leak-free nuclear umbrella, which Abrahamson thought no single system could do. Instead, he envisioned a series of layers, first targeting boost-phase missiles, then zapping warheads in space, and finally another layer close to the most crucial sites.[32] Add it all up and it began to look a bit like a Manhattan Project.

DARPA's directed-energy office moved over to the new agency. "My entire office literally just got moved out of the DARPA building and into the Washington building where SDIO was operating from," recalls office head Louis Marquet. They arrived to find a big picket line in front of their office. Star Wars was controversial, but the protestors were demonstrating against nuclear power.[33]

General "Abe," as he was known to one and all, retained his interest in the shuttle in his new job. The Pentagon was paying for part of the shuttle program, and getting a number of shuttle flights in return. The directed-energy office had not planned to use the shuttle, but after Abe suggested they use it, Marquet found three experiments that could benefit from a shuttle flight.[34] Abe knew nothing about lasers when he arrived, but Marquet gave him a crash course on an air force jet flying to California and was impressed by how quickly he learned. "Definitely the right guy at the right time for the program," said Marquet.[35]

Everything was in flux as SDIO geared up. After years of work, optics researchers were finding ways to clear the air by monitoring air turbulence and flows and adjusting the shape of mirrors continually to direct laser beams through it with minimal perturbation. That changed the rules, Abrahamson told a House committee. "Work with atmospheric compensation technologies has progressed to the point where it appears that the potential for large, effective ground-based lasers is very real."[36] Teller suggested popping mirrors up into space when needed.[37] ("I vastly prefer leaving lasers on the ground and sending up reflecting devices at the last possible moment," he wrote me in a letter.)

New technology offered promise for giant mountaintop lasers to send powerful beams through the air to space-based relay mirrors. John Madey argued the case for free-electron lasers, powered by massive electron accelerators, to feed high-energy laser beams to a network of relay mirrors that could relay beams to zap targets around the world. An alternative laser source for relay mirrors were giant versions of the ultraviolet excimer lasers powered by lightning-like controlled electric discharges. Star Wars also looked at ways to fire beams of electrons or other charged particles through space like supersized lightning bolts, although the technology was difficult.

The Star Wars program continued to support the Space Laser Triad mission, but questions were raised if the five-megawatt laser with a four-meter mirror would be adequate for missile defense. Some argued for a laser able to generate ten or twenty-five megawatts, and mirrors ten to fifteen meters in diameter, depending on the target. As a pointing-and-tracking system, Talon Gold would have required little change to work with large mirrors and higher laser power.

1985—SEARCHING FOR HELP

A year after SDIO was born, it was still sorting out how to achieve Reagan's goal of eliminating the threat of nuclear missiles. Its top managers traveled the country, making connections and holding meetings. I dropped in at one held April 18–19, 1985, at the University of Rochester, a center of optical technology since the nineteenth century. When I mentioned "Star Wars," a university administrator politely but firmly told me the proper name was the Strategic Defense Initiative. As a reporter I didn't have to be politically correct, but the scientists and engineers were much more careful because they were hustling SDI money.

The director of SDIO's directed-energy office made it clear he had his hands full. "We cannot answer the question 'will it work?' because we don't know what 'it' is," said Louis Marquet. He said strategic defense should prevent war; it would fail if it had to be used. He gave a laundry list of new technologies needed, mostly in space. The allure of lasers was their ability to deliver lethal energy at the speed of light, but they weren't up to the job yet. His goal was to understand technical issues well enough in the early 1990s to decide on going ahead with engineering development. That was ambitious, but far short of Wallop's crash program. As long as it was research, it was allowed by the ABM Treaty.[38]

Developers talked about rocket-engine, excimer, and free-electron lasers.[39] Dennis Matthews of Livermore had reported the first laboratory X-ray laser at an October 1984 meeting in Boston,[40] but none of the speakers said a word about the bomb-driven laser until someone asked keynote speaker George Keyworth about prospects for it.

"I think [it's] not likely. I say so categorically because it's very controversial," Keyworth answered. He noted that treaties limit the prospect for testing such weapons and said that a panel headed by James Fletcher to study prospects for SDI realized the importance of bomb-pumped X-ray lasers but did not think they were as useful as other proposed systems. He thought the United States should study them to know what was possible, and if the Soviets might make one, but he did not consider them a vital part of SDI, adding "in a democratic society, it's unlikely the public would welcome deploying [them]."[41]

DOUBTS IN THE LASER WORLD

A month later I attended the American laser industry's biggest meeting of the year, the Conference on Lasers and Electro-Optics in Baltimore. Star Wars was in the air in hallway discussions, but most were skeptical. It was the laser's twenty-fifth anniversary, and old-timers were out in force. Townes and Gould had both come out against Star Wars. The only people optimistic about laser weapons were those working at Livermore or the big aerospace companies. That didn't stop them from taking Star Wars money. They might denounce it as a waste of money, but as long as money was being thrown around, they were quite willing to hold out their hats and collect whatever fell in.

Defense contractors, in contrast, were veritably rubbing their hands in glee. A friend who worked for a small Boston-area company invited me to give a talk there about laser weapons later in November, so I shared what I knew. At lunch afterward, a company vice president was over the moon with the business prospects. With Reagan willing to spend billions, what could possibly go wrong?

I couldn't resist a subversive joke. "What happens if peace breaks out?"

I got the chuckles I expected, but none of us had much real hope. We were forty years into the Cold War. I, and many others around the table, couldn't remember a time when the United States and Soviet Union were not in a nuclear stalemate. We didn't expect much from the new leader of the Soviet Union, Mikhail Gorbachev.

INSIDE STAR WARS

A little-appreciated reality was that Star Wars was a research program in the very early stages of figuring out how to reach its goals. Reporters didn't understand that. The challenge of dealing with them, former SDI deputy director Gerold Yonas wrote recently, "was that they wanted to hear we had something ready to go, but we really didn't have any answers." He recalled telling reporters, "We can't make cost estimates until we do research. It will take five years until we know if we have anything worth developing. Until then it's a research project."[42]

Star Wars director Abrahamson was a three-star general who signed formal technical memos with a happy face. "Abe really believed in technological miracles, and he was able to convince others they could achieve more [than] they had expected," wrote Yonas. The general asked probing questions, but not antagonistically.[43] He seemed a perfect fit for the job.

Edward Teller may have been the biggest problem within Star Wars. He had spent more than four decades of his life on nuclear weapons and was often credited as the father of the hydrogen bomb. He had come to believe that the X-ray laser was an ultimate weapon, the third generation of nuclear weapons. The atomic bomb had been the first. The hydrogen bomb, called the "Super" in his day, was the second generation because it was so much more powerful than its predecessor. He saw the X-ray laser as a new generation because it focused its energy in a lightly directed beam. He could see it as a true way of taming the release of the energy contained in the atom in a controlled way. The X-ray beam was not a weapon of mass destruction, like the earlier generations; it was directional, a death ray that would destroy only a single target deserving of destruction. Perhaps Teller saw it as a way to redeem the damaged reputation of all things nuclear. It also was his way to defeat the Soviet missile threat.

"Teller was obsessed with the [bomb-driven X-ray laser] concept, even though the details were a bit fuzzy and left to his colleagues," wrote Yonas.[44] As a theorist, Teller tended to think the problem was solved once the concept was developed. After he had worked out an idea, he would claim it was ready for the engineers to handle the details.

One of the details that caused problems for Teller was selling the idea to Reagan, who wanted to rid the world of nuclear weapons. Many viewed Reagan as easy to manipulate; as mentioned before, his aides tried to keep Teller and others they saw as potentially manipulative away from the president. After a long effort to get through to Reagan, Teller finally arranged a meeting and went on at great length about how the nuclear-powered laser could defend against nuclear missiles. After he left, Reagan turned to an aide and mimicked Teller's Hungarian accent to say "Edvard loves da bomb."[45] Reagan was well aware of the nature of the X-ray laser and would have nothing of developing a new generation of nuclear weapons. Yonas supported research on it, at the behest of Nobel laureate Hans Bethe, to determine if the Soviets could build such a weapon because of its potential for use against SDI and its implications for national defense.[46]

Lowell Wood also realized the need to sell the bomb-driven X-ray laser and shrewdly invited *New York Times* reporter William J. Broad to stay at his house while Broad spent a week in May 1984 talking with the "Star Warriors" in Wood's group who were developing the X-ray laser. The result was a book titled *Star Warriors: A Penetrating Look into the Lives of the Young Scientists behind Our Space Age Weaponry.*[47]

The book portrays Wood's O Group, a group of bright, young, and mostly single white male physicists working on the bomb-driven X-ray laser. Most of their work was on computers because they couldn't just run down to the lab to test a tweak in their design. That required a nuclear test, which had been limited to being performed underground since the signing of the Partial Nuclear Test Ban Treaty in 1963, and they had to justify their project and wait for a slot in the schedule.[48]

Broad's main focus was on the personalities of the Star Warriors. I had spent four years as a Caltech undergraduate, at a time when most students majored in hard sciences like physics, astronomy, or engineering, so they seemed eerily familiar. In fact, one manager in the group, Tom Weaver, had lived across the hall from me for a year. We shared a taste in music, and I taped some of his albums, including the Rolling Stones singing "Sympathy for the Devil." The reel-to-reel tape is somewhere in my collection. Tom was one of the less boisterous of the denizens of the hallway we shared in Blacker House. He was a big, calm guy, and at the time he seemed the sanest, soberest, and most studious of us all.[49]

I lost track of him after graduation, but Broad's book filled in the details. Weaver had gone to Livermore, lured by a fellowship that let him work on his doctorate at the nearby University of California at Berkeley. His dissertation was on astrophysics, the role of shock waves in isotope synthesis during supernovas. His work was computer-intensive, and Livermore had some of the world's most powerful computers, a must to model the complex physics of nuclear weapons. Astrophysics is about the internal workings of stars, nebulae, galaxies, and all the rich variety of objects that glow in the universe. Supernovas are nature's nuclear explosions, far larger in scale than nuclear weapons, but sharing common physical properties. So many young astrophysicists drifted into working on bombs because jobs were easier to find than professorships.

Broad depicts the whole group as intense and energetic, most of them working long hours as they had in grad school.[50] Perhaps it was a way for them to stay young and engaged in an environment that in some ways paralleled a university, but one where the physics department and computer center were isolated far from the rest of the campus, surrounded by security gates, fences, and rules. But by 1985 when Broad's book was published, some Star Warriors were getting restless, and Peter Hagelstein would soon leave to join the MIT faculty.

THE *CHALLENGER* DISASTER

The space shuttle program got off to a slow start in 1981 but had managed a record nine launches in 1985. As the heavy launch vehicle of choice, it was important for Star Wars because it would be needed to carry heavy equipment into space for tests. Those plans came to a sudden end with the explosion of the space shuttle *Challenger* on the morning of January 28, 1986. Among the seven crew members lost was mission specialist Ronald McNair, a laser physicist with a PhD from MIT.

General Abrahamson was watching in his office with Marquet. It was the worst tragedy NASA had ever suffered, and it was televised live, the video replayed for all to see. It was a shattering experience for NASA, especially once the causes were revealed.

It was a bitter lesson for space enthusiasts; space was not as easy as they had hoped. Two years and eight months would pass before the shuttle flew again in September 1988, building up a backlog of large satellites waiting for launch. The Pentagon had to rethink its use of the shuttle and began shifting future launches to other vehicles. And Star Wars had to add access to space to its list of potential issues.

TWO MEN WHO HATED NUKES

Edward Teller did not fully appreciate that Ronald Reagan wanted not just to end the threat of nuclear war but also to rid the world of nuclear weapons. The Soviet establishment may not have realized that Mikhail Gorbachev likewise wanted to end the nuclear stalemate. But when the two men met in Reykjavik,

Iceland, for their first summit meeting in October 1986, they realized they shared a common dream, of abolishing nuclear weapons.

The two men agreed on their central goal but could not come to a full agreement in the conversation that followed. Gorbachev insisted that Star Wars had to "be confined to research and testing in the laboratories." Reagan said no. Again and again, Gorbachev pleaded for confining Star Wars to the laboratory. Again and again, Reagan refused. As the conference approached its last minutes, Reagan said, "It would be fine with me if we eliminate all nuclear weapons." "We can eliminate them," Gorbachev said. But then he repeated his demand that SDI stay in the lab, and Reagan repeated his refusal.[51]

They did not walk away empty-handed. In the early years of the Reagan administration, the United States and Soviet Union had talked about a ban on intermediate-range ballistic missiles, those with ranges below the minimum 5,500 kilometers of ICBMs. In practice, a major concern was Soviet missiles that could strike Western Europe, and US missiles stationed in Europe that could strike the Soviet Union. Before leaving Reykjavik, Reagan and Gorbachev agreed to remove intermediate-range missiles from Europe and limit their own arsenals to one hundred warheads each.[52] After further negotiations, Reagan and Gorbachev signed the treaty in December 1987 at a Washington summit, and destruction of warheads began. That began an era of nuclear disarmament that lasted until the 2000s, with thousands of warheads dismantled and large amounts of highly enriched bomb-grade uranium blended down for use in nuclear reactors.

At the time, it seemed a missed opportunity. The two leaders who shared a common vision of a better world without nuclear weapons had failed to come to an agreement because of disagreements that from a distance seem minor, almost terminological. What went wrong?

Looking back, Yonas says the reasons were complex. Neither of the two leaders understood the technological issues involved. Both had been getting conflicting advice on the status of strategic defense research. Former secretary of state George Schultz told Reagan that it would be worth trading SDI's limited accomplishments for a meaningful arms agreement. Yet Teller and others had claimed that Star Wars held the crown jewels of American technology. Some Gorbachev advisors told him neither SDI nor the Soviet Union's own efforts to develop strategic defense were getting anywhere; others boasted that a coming rocket launch would put the Soviet Union's space weapons into space for the first time. Some advisors warned that the Soviet Union could not afford to continue the Cold War, but others warned him not to surrender. Reagan saw the American economy as vigorous; Gorbachev could see the sickness of the Soviet economy.[53]

The odds were against success. "Neither of them had the backing from their supporters and assistants," wrote Yonas. Both had been getting disinformation. Both would have had to deal with enormous resistance from political and military factions whose thinking had been shaped by decades of nuclear stalemate.[54] Had the two finally succeeded in shaking hands in agreement to ban nuclear weapons, both would have faced difficult paths forward. The limited reductions they later agreed upon may have been the best they could have done.

POLYUS: THE RUSSIAN LASER LAUNCH—1987

US officials widely assumed that the Soviet Union had its own counterpart of the Star Wars program, but they did not know the details. In fact, at the time of the Reykjavik summit, Soviet plans were well underway for the first launch of its new giant Energia booster carrying a ninety-five-ton payload called Polyus that had been designed as a weapons platform for use in space (see photo insert). Energia's payload capacity was second only to the US Saturn V rockets that had launched Apollo on its way to the moon, and the United States had abandoned those costly giants for the space shuttle.

Soviet rocket engineers had designed two versions of Polyus, one called Skif that would carry a laser to target objects in low orbits or on ballistic tracks, and one called Kaskad, laden with kinetic interceptors that would target objects in higher and geosynchronous orbits. At the time of Reykjavik, the Russians had planned to launch a megawatt-class carbon dioxide rocket-engine laser, but after the summit they decided to load Polyus only with xenon and fluorine gas rather than carbon dioxide so the laser could not fire a beam. Later they decided not to release gas at all so the laser could not emit anything that could be interpreted as a weapon.[55]

All was ready when Gorbachev and other members of the Soviet politburo arrived at the launch site on May 11, 1987. Near the end of a tour of the site the next day, Gorbachev told engineers that they could not launch the rocket. However, after the engineers protested, Gorbachev and the other politburo members changed their minds and let the test proceed. The rocket successfully reached orbit, but before it was time to launch the Polyus module, the spacecraft rotated ninety degrees in the wrong direction, pointing Polyus on a downward course rather than an upward course into orbit. When the engine fired, Polyus separated and flew down to crash into the Pacific. It would be the only weapon-grade laser ever to reach space, although it was not functional and never reached orbit. US surveillance satellites recorded the activity but did not know what the Energia had carried.[56]

In 1988, a second Energia launched the unpiloted Buran, a shuttle-like reusable spacecraft that was only used once.[57]

BAD NEWS FROM NUCLEAR TESTS

Livermore conducted no nuclear tests of the X-ray laser between the late 1980 Dauphin test that indicated success and Reagan's Star Wars speech. The next test came just three days after the speech, but the instruments failed before they yielded any results. However, a December 16, 1983, test of Hagelstein's design yielded results that seemed to show laser action. Teller mailed Keyworth a letter declaring that this proved the scientific feasibility of the X-ray laser, and it was "entering engineering phase." That was too much for the more cautious Woodruff, who was responsible for the X-ray laser program. He demanded a retraction, and when Teller refused, he prepared his own letter, but Livermore's director Roger Batzel told him not to send it.[58] It was the start of a major split between Woodruff and Teller, who had tremendous clout despite being officially retired and having no management responsibility.

The tension grew after a Livermore measurement expert, George Maenchen, warned that the systems needed to verify the measurements had not been calibrated properly to identify the source of the X-rays they observed.[59] On August 2, 1984, the rival Los Alamos National Laboratory tried a test using different instruments but saw little or no X-ray lasing and concluded the X-ray laser did not exist. C. Paul Robinson, a high-ranking Los Alamos official, sent a caustic letter warning Livermore that they needed to get Teller and Wood to stop promoting it if Livermore wanted to keep its reputation.[60]

Meanwhile, Teller's claims escalated. On December 28, 1984, he wrote US arms negotiator Paul Nitze, "A single X-ray laser module the size of an executive desk which applied this . . . technology could potentially shoot down the entire Soviet land-based missile force if it were launched into the module's field-of-view."[61]

The basis of that claim was a new design by Lowell Wood's merry crew. The original Excalibur was supposed to concentrate the power from the bomb by a factor of a million by essentially optically pumping the ends of long, thin rods of X-ray-emitting material. In short order, Wood's theoreticians came up with two new variants—Excalibur-Plus that concentrated the power another factor of a thousand, and Super Excalibur that concentrated the power yet another factor of a thousand, to a trillion-fold enhancement. Excalibur-Plus supposedly was so powerful it could have zapped targets through the upper atmosphere if it was in low earth orbit or popped up from the ground. Super Excalibur could have struck such targets from geosynchronous orbit, about thirty-six thousand kilometers or twenty-two thousand miles above the ground.[62]

A Livermore graphic representing one version of Excalibur shows a sphere bristling with long, thin spines pointing in all directions, like a white porcupine curled up in a ball, hiding the Death Star of the bomb itself inside (see photo insert). Livermore's computer models predicted that the long, thin rods would convert X-rays from the bomb into tightly focused laser beams, but no experiment had verified that. The problem of how to point the multitude of laser beams at a multitude of targets moving in a multitude of directions was something Teller would have shrugged off as a mere engineering problem. An engineer would have known better and described it as somewhere between nightmarishly complex and utterly impossible. It was a pointing-and-tracking problem, which DARPA considered to be the hardest challenge of the Space Laser Triad. Broad later wrote, "Super Excalibur had no reality apart from a few paper studies. It was more hope than invention."[63]

That didn't stop Teller from hinting at SRI International that new technology would soon let America stop an attack by 5,000 missiles and 300,000 decoys. Wood briefed General Abrahamson on October 15, 1984, then spent much of the next year shuttling back and forth to Washington, telling federal officials how the new X-ray laser scheme would work.[64] Teller boasted about Super Excalibur in a December 28, 1984, letter to national security advisor Robert C. McFarlane and said it might be demonstrated in principle in as little as three years.[65] However, Wood was more restrained about timetables when talking to Livermore managers a month later, estimating that the scientific principles might not be established until 1990, even with generous funding. Not until then could they give the go-ahead to serious research and development and building a prototype, which might have a thousand beams.[66]

Wood's optimism grew after the March 23, 1985, Cottage nuclear test to validate Super Excalibur, which supposedly showed a million-fold increase in the beam brightness. The next month he told Central Intelligence Agency director William Casey that Super Excalibur would have "as many as 100,000 independently aim-able beams" with a glossy diagram, adding that a nuclear test in March had shown the beam could be focused to weapon-level brightness.[67] Teller and Wood's campaigns seemed to be on a roll, with big-budget increases projected in coming years (see photo insert).[68]

But problems were beginning to surface. In November 1985, *Science* magazine reported that Los Alamos had found problems in the Cottage test that made Livermore greatly reduce its estimates of X-ray brightness. The same article also raised doubts that bomb-driven lasers could be the ultimate weapon to foil massive nuclear attacks. If they were stationed in orbit, they would be vulnerable to antisatellite weapons that could disable the X-ray laser before an attacker launched a first strike. Wood and Teller's pop-up defense had its own fatal weakness; it relied on detecting a massive enemy launch almost instantly, which the enemy could overcome by building boosters that burned much faster than conventional types to reduce warning time. "In the end, the pop-up x-ray laser is simply not feasible against a fast-burn booster," said Cory Coll, director of SDI systems analysis at Livermore. "Fast-burn boosters rule out pop-up anything."[69] The Goldstone test at the end of December, planned before the story broke, did even more damage. When all the numbers were crunched, the X-ray laser was only one-tenth as bright as Livermore had thought.[70]

At the end of October, Woodruff resigned as associate director of the nuclear weapons program, frustrated by top management's failure to control excessive claims made by Wood and Teller, who nominally reported to him.[71] Hagelstein, whose models had helped shape the bomb-driven X-ray laser, left to take a faculty job at MIT.[72]

It was a bumpy road from there, mostly downhill. The next round of tests sought evidence that the X-ray beams could be focused to concentrate their energy, which had been unclear from the December Goldstone test. The Labquark test in September 1986 looked encouraging, but more sophisticated instruments in the next test, Delamar, revealed bad news. The beam wasn't shrinking in size, as if it were focused; it was spreading out in a ring that left a hollow central region of very low X-ray intensity.[73] That did not prove that X-rays could not be focused to high intensity levels, but it showed that Livermore hadn't done it. The lab's elegant theoretical models did not match reality. SDIO started phasing down the X-ray laser budget.

The scientific fiasco blew up into a scandal when an unknown whistleblower at the University of California, which ran Livermore, sent some personnel files to the Southern California Federation of American Scientists. Livermore's management had shunned Roy Woodruff after rejecting his protests about Wood and Teller. He had hoped to move to another program at the lab, where he had worked since college. Instead he was relegated to a small office, which his friends called "Gorky West" after the Russian city Gorky where the Soviets exiled dissident scientists. Woodruff had appealed through channels, and the whistleblower found those files. The Federation of American Scientists group released them in October 1987, and the news soon made headlines in California.[74]

That was something Woodruff had not wanted. He was a firm believer in nuclear weapons and a strong defense, but he also insisted that everyone played by the rules and told the truth. Woodruff was at Los Alamos when the news broke, and scientists there gave him a standing ovation. The General Accounting Office found that no laws had been broken and expressed some concern about the process of dealing with Woodruff's complaints. They said Teller and Wood were far more optimistic about the X-ray laser than anyone else.[75] Teller and Wood kept their posts at Livermore, but the lab took a serious hit to its credibility.

Teller and Wood's credibility suffered more. Broad, whose 1985 book *Star Warriors* cast a generally favorable light on Wood and the young physicists working on the X-ray laser, dug deeply into the collapse of the X-ray laser. In 1992, he followed up with a harsh but well-documented book covering the fiasco, *Teller's War: The Top-Secret Story behind the Star Wars Deception*. His prologue laid out the results of Teller's quest for what he had hoped would be his final success, an ultimate weapon, although Broad does not use the word. "The result, however, was no triumph. Over the protests of colleagues, Teller misled the highest officials of the United States government on a crucial issue of national security, paving the way for a multibillion-dollar deception in which a dream of peace concealed the most dangerous military program of all time."[76]

Others who felt Teller and Wood had deceived them were also harsh. George Keyworth, Reagan's first scientific advisor, later called the push to incorporate bomb-driven X-ray lasers into the Star Wars program "a pack of lies, unadulterated lies."[77]

THE QUEST FOR NEW LASERS

At the end of 1988, Star Wars was developing several types of laser weapons, Richard L. Gullickson reported at the Lasers '88 conference in Lake Tahoe, Nevada. MIRACL, now the center of a national laser weapon test range, was being used for ground-based tests with a beam director to test the lethality of high-energy lasers on various targets. TRW was testing Alpha at its facility in the hills east of San Juan Capistrano, California. Another variation on the hydrogen fluoride rocket-engine laser was being tested at half of the hydrogen fluoride wavelengths, but that project would fade away.[78]

A different approach was building giant mountaintop lasers that could focus their beams up to a relay in orbit. Free-electron lasers seemed to be the best option for a mountaintop laser. They had to be big because they were built around massive accelerators that produced relativistic electrons that amplified the light in the laser beam. As of late 1988, free-electron laser powers had only reached the watt level in the laboratory, but that didn't stop planners from envisioning that the planned Ground Based Free Electron Laser Technology Experiment (GBFEL-TIE) could generate a megawatt beam in the mid-1990s.[79] A ground survey was seeking a site two miles wide and ten miles long (3.2 x 16.1 kilometers) on the White Sands Missile Range.[80] At the time, some considered free-electron lasers to be the most promising approach, Gary R. Goldstein told a 1988 conference of the American Physical Society. But his analysis found "at best, future facilities will be many orders of magnitude away from the required gigawatt average output powers in the visible or near infrared region" needed for nuclear missile defense.[81]

Those mountaintop lasers would never be built. In May 1989, SDIO pushed plans for building them for missile defense out beyond the year 2000.[82] Another budgetary whack came the following year, ending any plans for major construction projects.[83]

SPACE-BASED LASERS

Progress on the Space Laser Triad was slow under SDI. Its marquee project was the Alpha laser, officially rated at megawatt class. At one point it was planned to generate five megawatts, but it never reached that level. Alpha was superficially similar to the two-megawatt MIRACL rocket-engine laser, in the sense that it generated megawatt-class power by burning chemical fuels to produce molecules that emitted infrared light. The molecules were almost identical but contained different isotopes of hydrogen. MIRACL produced hydrogen fluoride containing the rare heavy isotope deuterium because it was built to run in air, which absorbed relatively little light at its wavelength of 3.7 micrometers. Alpha produced hydrogen fluoride containing the abundant isotope hydrogen-1 because space lasers would not be affected by air absorbing its 2.8-micrometer wavelength.

But there the similarity stopped. MIRACL was designed in the 1970s to test the prospect of putting big lasers on US Navy battleships to zap enemy cruise missiles, a major concern at the time. Battleships are big enough that minimizing size and weight were not big concerns. However, the rocket-engine chemical laser could not run efficiently at high pressure, so little of the energy from the chemical reaction was converted into laser light, and MIRACL was building-sized with the beam emerging from a modest area.

Alpha had to meet different requirements because it was a test of feasibility of lasers based in space, which would have to cope with different conditions. One important advantage of space-based lasers is that they would operate at low pressure, which allows rocket-engine lasers to run much more efficiently than MIRACL could at atmospheric pressure. However, a new laser design was needed for low-pressure operation, and a special facility was needed to house it, so a giant white structure grew over the hills east of San Juan Capistrano to house the Alpha laser facility.

The body of Alpha was a cylinder about a meter in diameter with laser exhaust flowing radially outward from twenty-five rings stacked vertically. Laser light emerged at right angles to the rings, going vertically along the outside of the cylinder. Mirrors at the top and bottom of the cylinder reflected and collected the laser light and formed it into a powerful beam.[84] The laser first operated on April 9, 1989, and within a couple of years it had reached megawatt-class output. Developers reported in 1995 that they routinely produced megawatt-class output with excellent beam quality in runs lasting up to 6.2 seconds.[85]

A second key element was a four-meter segmented mirror designed and built by Itek for use in space, which was featured in the November 23, 1987, *Aviation Week* and displayed on its cover. Ed Gerry told the House Armed Services Committee that the four-meter mirror had "exquisite optical quality and . . . the ability to adjust the surface as necessary to remove distortions." Martin Marietta had just been awarded a $108 million contract to test the giant laser and the giant mirror as a part of a study called Zenith Star to assess prospects for combining the two—or versions of the equipment usable in space—for flight experiments. "Technologies required for a near-term space-based chemical laser system study are rapidly approaching maturity," Gerry said, and he predicted that a militarily usable laser weapon system suitable for boost-phase interception and midcourse interactive discrimination "appears to be within reach before the end of the century."[86]

President Reagan visited Martin Marietta in November 1987, where he appeared onstage with a mock-up of Zenith Star, an eighty-foot-long, fifteen-foot-wide cylinder (see photo insert). It was so big that launching it would require a massive new cluster booster with eight to ten million pounds of liftoff thrust, more than achieved by either the Soviet Energia or the US Saturn V.

The planned booster was so gargantuan that it was called variously "the Barbarian" and "Huey" for humongous.[87] There was talk of scaling up laser powers to as high as twenty-five megawatts and building segmented mirrors that would unfold to sizes as large as fifteen meters so the laser battle stations could destroy nuclear missiles.

The planned test would see if the Zenith Star system could control vibration and shape the beam so it would remain on target a hundred to two hundred kilometers away. Star Wars officials said the laser didn't have to destroy the target to meet their goals, just stay on target. Yet although officials sounded optimistic, the schedule for the Zenith Star test soon started slipping. In late 1987, officials talked of tests in space by 1990.[88] A year later, Zenith Star had grown into a $1.5 billion project that would span seven years and would start tests in mid-1994.[89] A 1989 report from the General Accounting Office reported "as of March 1989, the launch date had slipped from 1990 to the mid-1990s."[90]

With the Soviet strategic missile threat fading, interest and budgets shrank as well. TRW reported encouraging results in tests of the feasibility of integrating such a large mirror with Alpha.[91] A full-scale ground demonstration followed in 1997.[92] But the space-based laser and Zenith Star would never get off the ground. Nobody wanted to pay to build the gargantuan booster, so Zenith Star was scaled back to two modules that would be assembled in space. Eventually the project was renamed and downscaled to a ground-based test in 2008 before the plug was pulled in 2002. By then the big laser bet in ballistic missile defense had shifted to a new type of megawatt-class chemical laser squeezed into a Boeing 747, the Airborne Laser, covered in the next chapter.

BUSH ADMINISTRATION CHANGES THE RULES

The rules changed when Reagan left the White House. Under George H. W. Bush, the Pentagon officially recognized what technologists had been saying since the start of Star Wars: it was impossible to protect the whole United States from nuclear attack.[93]

Abrahamson left office, recommending with great enthusiasm the latest idea touted by Wood and Teller. They called it "Brilliant Pebbles," but the idea having high-speed interceptors simply homing in and crashing into nuclear missiles had been kicking around since the 1950s. New electronic technology made it more feasible in the 1980s, and Los Alamos physicist Gregory Canavan had talked with Teller and Wood in 1986 about his ideas of using "smart rocks" for missile defense. Wood had taken careful notes and advanced the design to become Brilliant Pebbles.[94] One commentator called the plan "loose marbles," but SDI didn't have many options left.[95] Spending on the new plan ramped up, and advocates of longer-term research complained, "Brilliant Pebbles is eating every other programs' lunch."[96]

American officials opening arms control negotiations showed Soviet visitors the Alpha rocket-engine laser at TRW in San Juan Capistrano in 1989. On June 22, 1991, SDIO showed the facility to the American public for the first time, after a May 16 test had exceeded a megawatt, a feat the head of SDIO's directed-energy group said "shows it's real."[97]

The power came from the Alpha rocket-engine laser, a cylinder 1.1 meters in diameter and two meters long, the size of an overweight refrigerator. Planners claimed that a five-meter-long version could produce about ten megawatts. The rocket-engine laser itself stacked together twenty-five identical hollow aluminum rings that sprayed nitrogen trifluoride and hydrogen gas inward where they burned to form free fluorine that was accelerated outward through nozzles to supersonic speed and mixed with hydrogen to form hydrogen fluoride molecules that emitted laser light in a zone around the cylinder. The size of the vacuum chamber limited the power, and chemical scrubbers removed toxic hydrogen fluoride and fluorine. The next step was to be the laser with a four-meter optical system.[98]

However, by then SDIO was talking about space-based lasers as a follow-on to a much downscaled version of Brilliant Pebbles, proposed by new Star Wars head Henry F. Cooper. The world was changing as the Bush administration watched. The Iron Curtain had fallen, and the Soviet Union had begun to splinter. The first Gulf War had highlighted a new reality, where the most immediate threat was that "rogue states" might launch missiles at US allies or perhaps the United States itself. Cooper called his plan GPALS, for Global Protection Against Limited Strikes, and envisioned building a thousand Brilliant Pebbles projectiles and five hundred to a thousand ground-based interceptors.[99]

That proposal got an unusually caustic reaction from *Aviation Week*. An editorial titled "Bring SDI Down to Earth" asked, "How much have Americans actually advanced the cause of national security for the $20 billion plus expended on SDI since 1983? Sad to say, the answer is not nearly enough." The editors added, "With a straight face, Henry F. Cooper, the SDI director, said the [$46 billion] cost is so *low* [italics theirs] because of the capabilities of Brilliant Pebbles." Even the traditionally pro-defense magazine had had enough of costly and ineffective space-based defense schemes.[100]

They weren't the only ones disenchanted. A week after the TRW Alpha facility was opened to the press in 1991, I was in San Juan Capistrano to cover a conference on near-Earth asteroids. When a Livermore speaker proposed a costly system to search for asteroids, an astronomer in the audience asked sarcastically if he was proposing "the Strategic Asteroid Defense Initiative." Astronomers were not happy about Lowell Wood giving a dinner talk.

THE END OF SDIO

The Soviet Union fell apart at the end of 1991. In February 1992, a report from the General Accounting Office warned that a series of design changes caused by the shift to the Brilliant Pebbles approach were creating large risks.[101] Later that year and into 1993 more bad news emerged. Test results for Brilliant Pebbles and other programs—including the key 1984 Homing Overlay Experiment—had been rigged or exaggerated. Data on Patriot missile kills had been fudged.[102]

In May 1993, the new Clinton administration erased the Strategic Defense Initiative name and turned a shrunken missile defense program over to the new Ballistic Missile Defense Organization. The Star Wars wars were over.

WHAT WAS STAR WARS?

Ronald Reagan started the Strategic Defense Initiative seeking new technology to produce an ultimate weapon that would protect America from nuclear weapons. His offer to share Star Wars technology with other nations reflected his hope that some such ultimate weapon was possible and would rid the world of nukes. Yet he and Gorbachev never managed to make the deal they both wanted. If they somehow had succeeded, the powers behind them might well have stepped in to stop the deal, blocking it in the Senate, the politburo, or some other back room.

Lowell Wood claims that Star Wars was a brilliantly successful bluff, a Pentagon version of a Potemkin village that displayed the illusion of an awesome technological capability, fooling the Soviets into essentially giving up. He says he knew all along that the idea he suggested was too complex and expensive to be practical. "I went into it with my eyes wide open, and I did the job. . . . I got the result that I wanted. The Soviet Union collapsed."[103]

Gerold Yonas, who spent two years as Star Wars chief scientist and wrote a book titled *Death Rays and Delusions* about his experience, disagrees.[104] "Nobody was bluffing. Reality was not so simple as a bluff. Reagan hated nukes and the Soviet Union. He believed we could defend ourselves if we could jointly manage a transition to eliminate nukes and share a defense system," he wrote in an email. "SDI had little impact and the Soviet Union went bankrupt on its own without our help. Deceit, mismanagement and moral confusion destroyed them."[105]

If it was a science fiction story, nobody would believe it.

An MIT course 2.009 project shows how Archimedes could have ignited Roman ships with a solar death ray, on the roof of a campus garage. The wood ignited less than ten minutes after the sun emerged from behind a cloud. (Photo courtesy of MIT course 2.009.)

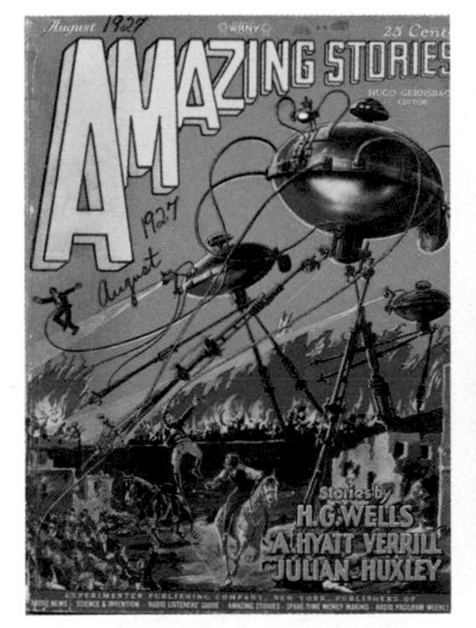

In *The War of the Worlds*, H. G. Wells described the death rays wielded by Martian invaders as invisible rays of heat that ignited anything they touched. But invisible rays weren't scary enough for the cover of *Amazing Stories* when it reprinted the novel in August 1927. (Image used with the acknowledgment of the Frank R. Paul Estate.)

Death rays come to the rescue when insectoid visitors from the moon encounter a hungry *Tyrannosaurus rex* in "The Death of the Moon" by Alexander Phillips. Cover of the February 1929 *Amazing Stories* painted by Frank R. Paul, one of the great artists of the pulp era. (Image used with the acknowledgment of the Frank R. Paul Estate.)

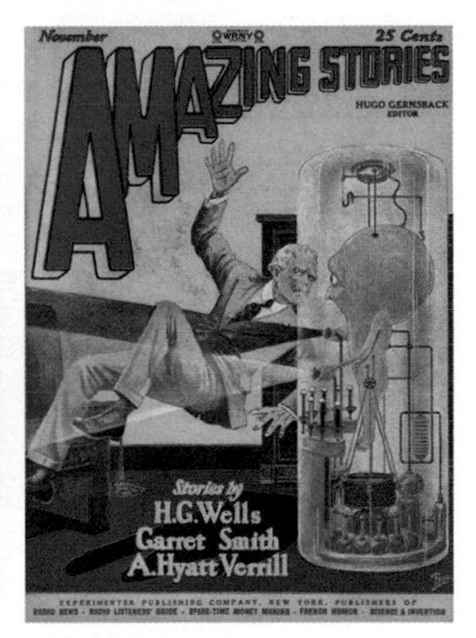

Not all science fiction rays were deadly. A man from thirty-five thousand years in the future uses manipulative rays to hold a man from 1927 in midair in a scene from "The Machine Man from Ardathia" on the cover of the November 1927 *Amazing Stories*, painted by Frank R. Paul. (Image used with the acknowledgment of the Frank R. Paul Estate.)

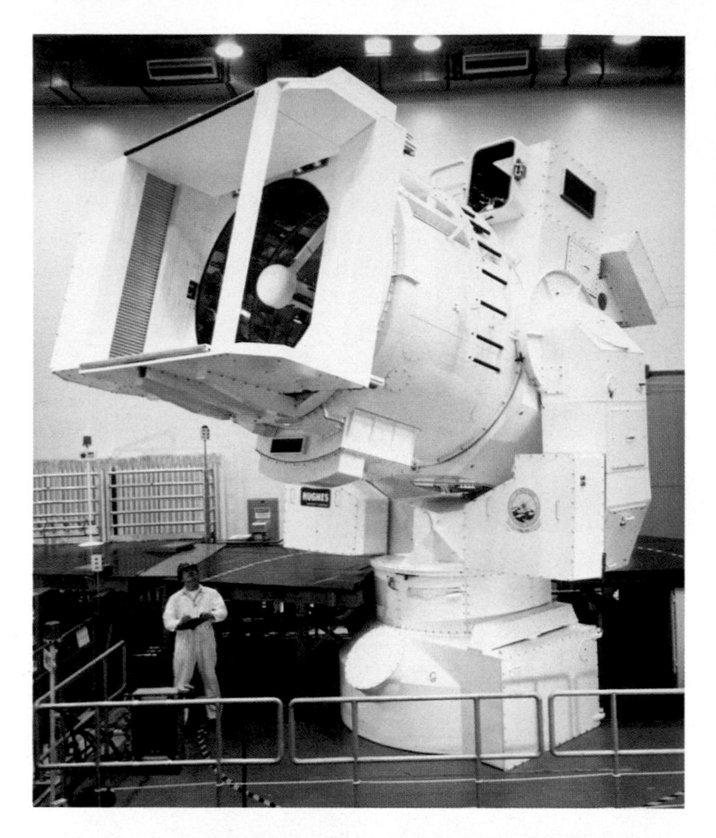

The Sea Lite Beam Director built to handle the two-megawatt output of the Mid-InfraRed Advanced Chemical Laser (MIRACL), shown at the factory. (Photo courtesy of the National Archives.)

Inside a rocket-engine laser is a lot of plumbing. Gases flow into the yellow-orange structure and mix to form hot deuterium fluoride, which flows through nozzles and emits two megawatts of infrared light. This is the Mid-InfraRed Advanced Chemical Laser, dubbed MIRACL. (Photo courtesy of the National Archives.)

Gary Tisone (*left*) and A. K. Hays (*right*) work on an early pulsed ultraviolet laser. Electron beams zapped gases inside the large tube to produce laser light from argon fluoride, a laser now used in eye surgery and making semiconductor electronics. (Photo courtesy of Sandia National Laboratories.)

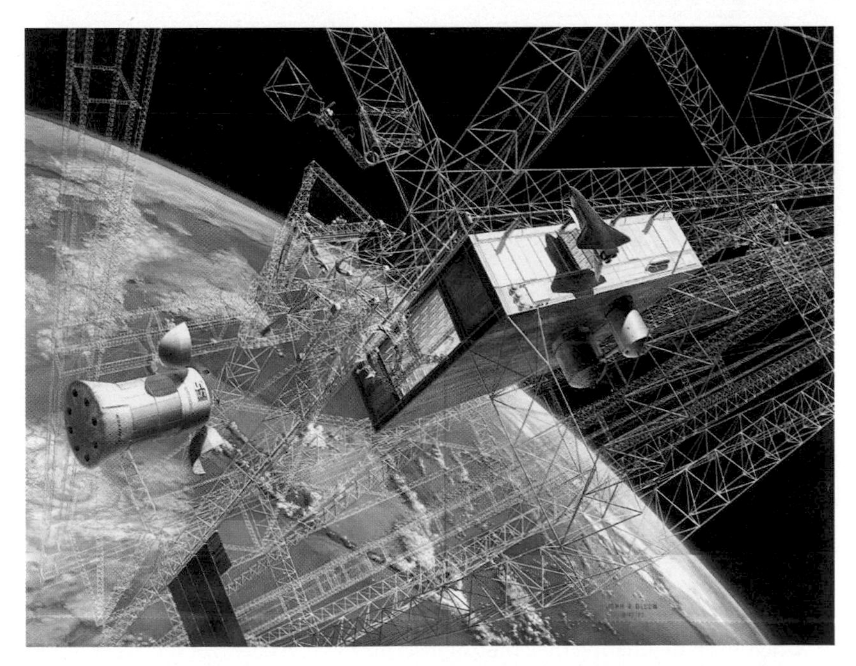

Assembly of a solar-power satellite in space, one example of how industry could be moved into space in the visions of Gerard O'Neill. Instead of burning polluting fossil fuels or paving open space on the ground with solar cells, power satellites in geosynchronous orbit could beam microwave energy down to small areas on the ground. (Image courtesy of NASA.)

The giant Soviet Energia rocket loaded with the black Polyus module carrying a prototype high-energy carbon dioxide laser unable to produce a laser beam in space. The May 1987 launch was intended to show that the Soviets could put a massive laser in orbit, but when the Energia reached space, the spacecraft rotated ninety degrees in the wrong direction and sent Polyus to crash into the Pacific. (Photo courtesy of Vadim Lukashavich, Buran.ru.)

A Livermore artist's concept of Super Excalibur, a version of the bomb-driven X-ray laser that resembles a cross between a porcupine and the Death Star. The white spines projecting outward are rods of X-ray laser material, which would be aimed at many individual enemy nuclear missiles. The burst of X-rays from detonating the bomb at its center would energize the X-ray laser material, producing a powerful laser pulse directed along the lengths of the rods and focused toward the points where the targets would be when the X-ray pulse reached them. In theory, a single Super Excalibur could zap large numbers of nuclear missiles in a single blast. However, tests failed to verify the model. (Image courtesy of the Lawrence Livermore National Laboratory.)

Edward Teller (*left, in pale jacket*) and Lowell Wood (*left, bearded*) brief the SDI brain trust. Ronald Reagan faces away from the camera and toward Teller. General Abrahamson is in uniform at right. The black cloth probably covers a model of a secret weapon envisioned by Teller and Wood. (Photo courtesy of the White House Photographic Collection, Reagan Library.)

Ronald Reagan speaking at a Martin Marietta plant in front of a mock-up of the massive Zenith Star spacecraft, which was planned for a space-based test of the high-energy laser and optics developed for the Space Laser Triad. Work continued through the 1990s, but the program was canceled in 2002. (Photo courtesy of the White House Photographic Collection, Reagan Library.)

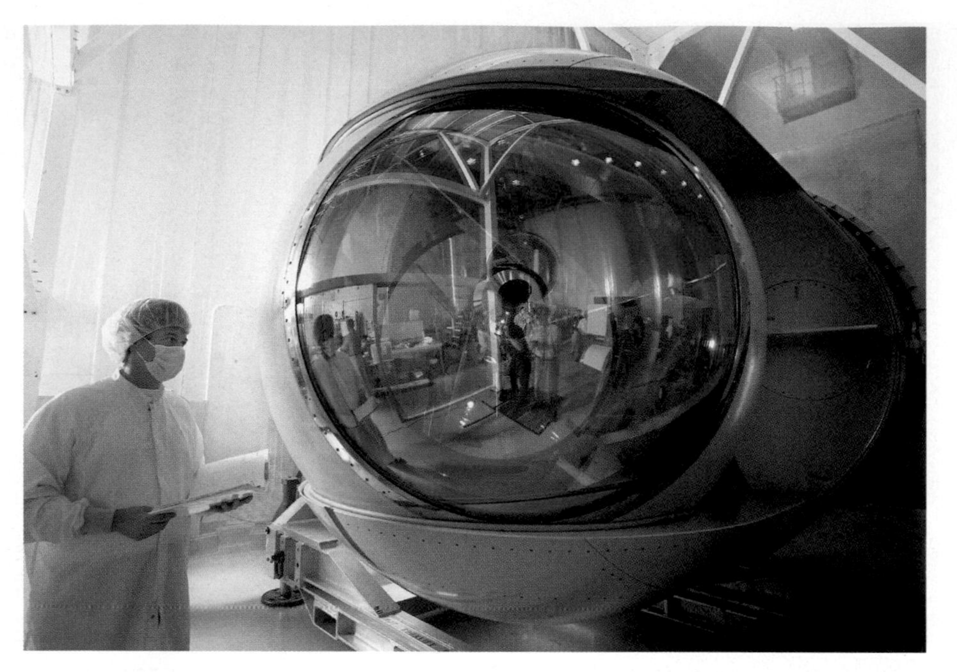

Close-up of the optical turret of the Airborne Laser after assembly. This was the final stage in a complex optical system that directed the megawatt-class beam toward its targets. It shows reflections from the lab overlaid on the 1.5-meter mirror that sent the laser beam on its way. (Photo courtesy of Lockheed Martin.)

THEL beam director mounted atop the trailers containing the rocket-engine laser and other equipment during tests at White Sands. Soldiers give the scale. (Photo courtesy of Northrop Grumman.)

The US Navy LaWS system looks over the ocean from the USS *Ponce*. Six industrial fiber lasers from IPG Photonics supply enough laser power to shoot down drones and detonate munitions on small boats. (Photo courtesy of the Office of Naval Research.)

The Marine Ground-Based Air Defense system: a laser on a combat vehicle zapping drones or missiles, a concept being investigated by the Office of Naval Research for the US Marines. (Image courtesy of the US Navy.)

Lockheed's Athena laser weapon system: a laser weapon that can fit on a truck, with the top removed to show the details. It can burn a hole in the hood of a pickup truck, as shown in the next image. (Photo courtesy of Lockheed Martin.)

A smoking-hot hole just burned in the hood of a pickup truck by Lockheed's Athena laser. The truck seems to have been tipped to make a better target, but the effect of the laser is still impressive. (Photo courtesy of Lockheed Martin.)

CHAPTER 6

THE AIRBORNE LASER GETS OFF THE GROUND

The end of the Cold War left the missile defense program in the awkward position of having lost its primary enemy. Star Wars faded away. But the first Gulf War in 1991 brought back one of the nightmare scenarios of World War II. Iraq launched short-range Scud ballistic missiles to bomb Israel and Saudi Arabia, the first use of such missiles in combat since the Nazis fired their last V2 rockets toward England. Scuds lacked pinpoint accuracy, but they were nonetheless deadly. The United States had Patriot missiles, but their success in defending Israeli and Saudi sites from the Iraqi attacks had been limited. That renewed US interest in an airborne antimissile laser.[1]

The Airborne Laser Laboratory had shown that a laser could shoot down flying targets from a plane in principle, but the first-generation rocket-engine laser technology was not ready for prime time on the battlefield, and its ten-micrometer wavelength was a problem. When former air force secretary Hans Mark took a fresh look at laser technology in the early 1990s, he saw two big advances that promised big improvements in antimissile lasers. One was the emergence of a new type of high-energy laser emitting at 1.3 micrometers, where air is more transparent than at the ten micrometers of the Airborne Laser Laboratory. The other was major improvements in adaptive optics, a technology that could compensate for air turbulence and beam dispersion.[2]

The 1.3-micrometer laser is called the chemical oxygen iodine laser (COIL) because it's based on chemical reactions transferring energy from excited oxygen molecules to iodine atoms that emit laser light at 1.315 micrometers. The process is quite complex and evolved over a number of years before being first demonstrated at the Air Force Weapons Laboratory in New Mexico.[3]

The chemistry starts with mixing a strong base like sodium hydroxide with concentrated hydrogen peroxide, producing a highly energetic free radical with two oxygen atoms, one hydrogen and an extra electron. That then reacts with molecular chlorine to form excited oxygen molecules that can transfer energy to iodine atoms.[4]

The chemistry is powerful, and it involves rapidly flowing gases, but it's not burning chemical fuels in the same way as other types of rocket-engine lasers.

The first COIL at the weapons lab in 1978 emitted only four milliwatts, and its efficiency was very low.[5] But the following year the same research team stepped the power up to one hundred watts with 3.7 percent efficiency, and by 1989 the power had reached thirty-nine kilowatts with 24 percent efficiency.[6] That was approaching weapon-like powers, and the wavelength might travel a hundred kilometers or more through air to zap missiles rising from the atmosphere. The technology looked mature enough that when Congress in 1992 told SDI to drop projects that could not be brought to fruition within fifteen years, a plan to develop an airborne laser using COIL technology easily made the cut.[7]

THE EMERGENCE OF ADAPTIVE OPTICS

A second, and equally crucial, factor was the development of new optical systems to reduce the impact of air powerful laser beams. The difficulty of delivering a lethal dose of laser energy through the atmosphere to a moving target became a showstopper for the first generation of battlefield laser weapon tests. Moisture and carbon dioxide absorbed part of the laser energy and warmed the air so the beam spread like a searchlight. Natural air currents made laser beams waver as if they were passing over a blacktop parking lot on a sunny day. Trying to track and hit targets was even harder when the Airborne Laser Laboratory was flying. The laser weapon suffered both from vibration of the aircraft and from the turbulence of air flowing around it. The beam could hit the broad side of a barn with the plane resting firmly on the ground, but in the air the Airborne Laser Laboratory struggled to keep the beam focused onto a moving target long enough to do lethal damage. The Airborne Laser Laboratory finally scored a few hits in 1983 after years of effort, but it was clear that the technology had to go back to the drawing board before there would be much hope for a practical battlefield weapon.

Astronomers had long had similar problems. Turbulence constantly jiggles starlight as it passes through the air, making stars twinkle at night. Photographs and electronic imaging can record images of extremely faint objects during long exposures in still air, but air turbulence spreads stellar pinpoints into fuzzy blurs in the night sky. In the 1950s, astronomer Horace Babcock had the good fortune to work at the Mount Wilson and Palomar Observatory, home to the world's two largest telescopes, the one-hundred-inch Mount Wilson telescope completed in 1917 and the two-hundred-inch Palomar telescope finished in 1948. With sensitive photographic film, the Mount Wilson telescope had revealed that the universe was full of distant galaxies. But atmospheric effects blurred the images so they were only as sharp as a ten-inch telescope could record. The Mount Wilson telescope is four inches bigger than Hubble, so without the air (and the lights from Los Angeles), in theory it could see the stars as clearly as Hubble. Frustrated with the blurry images, Babcock proposed a way to clear the sky. He envisioned using sensitive optical instruments to measure how air turbulence blurred starlight, and finding a way to use that information to continually reshape telescope optics to cancel out the blurring effects.[8] It was a brilliant idea, but the bulky optical sensors, massive vacuum-tube computers, and heavy glass telescope mirrors of the day were totally inadequate for the task. Nor did astronomers have the money needed to develop the new technology.

The Pentagon had plenty of money to invest in new technology to improve images of military interest. They tested ways to clean up photographs from spy satellites in the 1960s. In the early 1970s, ARPA sought a way to sharpen the view military ground-based telescopes had of orbiting Soviet satellites, a problem similar to that of astronomers. In 1972, ARPA funded the Itek Corporation along Boston's Route 128 to take a fresh look at Babcock's idea. Itek tried to measure how much air turbulence distorted the view of satellites from the ground. Then they tried to use that information to change the shape of a flexible "rubber" mirror so a ground-based telescope could show Soviet satellites more clearly than ever before.[9]

That required building an adaptive optical mirror with an array of small pistons behind a thin, flexible mirror surface. When sophisticated sensors measured distortion of the incoming light, controllers moved the pistons back and forth to reshape the mirror so it would show a sharper, more stable image. In 1974, Itek reported success, and *Laser Focus* ran a photo of a small rubber mirror on its December cover. It was an important first, with potential uses in getting images and optical communications through the air as well as high-energy laser beams. Itek claimed the mirror could overcome image distortion.[10]

The technology also sharpened the beam from a high-powered rocket-engine laser. Pratt & Whitney had built its Experimental Laser Device (XLD) at a plant in steamy southern Florida.[11] The high humidity soaked up that laser's ten-micrometer infrared light, heating the air in the center of the beam, causing the beam to spread outward, an effect called thermal blooming. The company had laid two miles (about three kilometers) of railroad track in a marshy area behind its plant so engineers could measure beam transmission. Louis Marquet from the MIT Lincoln Laboratory added a rubber mirror to the laser optics and installed a set of sensors on a railcar to be moved along the tracks to measure what happened. He found the rubber mirror system reduced spread of the laser beam up to 60 percent. There had been hope for more, but it still was a big step for adaptive optics.[12]

Meanwhile, DARPA's project to develop large optics for orbiting laser battle stations was facing the problem of how to haul big mirrors into space and maintain the extremely precise shape needed to zap distant targets. They decided the solution was to make large segmented mirrors that could be unfolded in space. By using thin mirror segments, they could both save weight and make the segments flexible enough to be adjusted by actuators like those used on rubber mirrors. As described in chapter 4, DARPA used that technology to build the four-meter mirror in the Large Optics Demonstration Experiment.

The Pentagon recognized adaptive optics was a crucial technology even as the armed services pulled back from laser weapons on the ground. By 1982, military researchers had doubled or tripled the number of pistons moving the surface, and increased the size of the rubber mirrors by four- to fivefold.[13] As the technology got better, the Pentagon worried more about security and tried to keep the details undercover. When security officials ordered some 120 research papers pulled from an open optics conference in San Diego in August 1982, many of the removed papers covered applied optics.[14]

By that time, the Air Force Weapons Laboratory had become a key center in the highly sensitive field of applied optics. Janet Fender, now chief scientist for the Air Combat Command, recalls a "can-do" culture when she arrived there in 1981. Although the Airborne Laser Laboratory had yet to shoot down a target while it was flying, the developers knew how far they had come from where they had started.

Like many optics specialists, Fender began with an interest in astronomy. She got a job doing astronomy at the Kitt Peak National Observatory but found she was more interested in working on optical instruments. That led her to the nearby University of Arizona, then one of only two universities in the United States that granted doctorates in optics. From there it was a short move to the weapons lab at Kirtland Air Force Base to work on adaptive optics and the cutting edge of military optics. Two years after her arrival, the Airborne Laser Laboratory's success shooting down Sidewinders and cruise missiles while flying was good news to the weapons lab, she says, because reports on the national news showed that the job could be done.[15]

Other projects kept the weapons lab optics group busy in the 1980s. The large optics being developed for space-based lasers needed adaptive optics to maintain their precise shapes for pinpoint beam delivery. Anything going into space must be lightweight, making it vulnerable to bending if any force is applied to it. A big concern in space is the temperature changes that come when satellites move into and out of sunlight as they circle Earth. NASA learned this the hard way when soon after its launch the Hubble Space Telescope started to shake each time it moved from light to dark because poles holding its solar arrays were shrinking and expanding.[16] When SDI decided to develop segmented mirrors that unfolded in space like umbrellas, those deployable mirrors also required adaptive optics to keep the segments properly aligned under stresses that might change in orbit. SDI funded two companies to develop the technology, hoping to develop an industrial base. Later the technology was made available to astronomers, and it was used to make the folding six-meter mirror on the James Webb Space Telescope.[17]

Other weapons lab projects explored other tools the scientists hoped might be useful in laser weapons. Fender and a colleague set out to combine beams of laser light from an array of telescopes to produce a single powerful beam with the coherence of light from a single laser. It was a concept already used in microwave radar, where an array of small antennas could be pointed in the same direction and their beams combined to make a single coherent beam focused in a beam much narrower than the wide beam from an ordinary radar antenna. But it was much trickier to do with light because the waves had to be matched up to within a fraction of a wavelength, which was centimeters for radar but tens of thousands of times smaller for lasers. Using deformable mirrors, they split light from a single laser among three parallel telescopes, then successfully combined the outputs into a single beam as coherent as the original laser.[18] They called their demonstration Phasar, a name inspired by *Star Trek* rather than *Star Wars*.

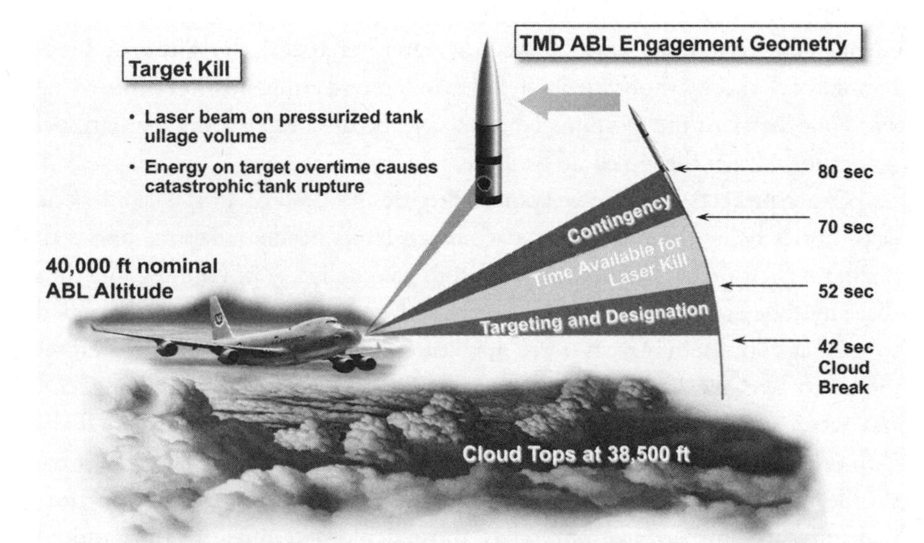

Figure 6.1. The Airborne Laser attack plan: The plane was to fly at forty thousand feet, just above the cloud tops. A ballistic missile needs about forty-two seconds to emerge from the cloud banks just below the patrol zone. The Airborne Laser would have ten seconds to target the missile before zapping it for the eighteen seconds needed for a laser kill. The laser could continue focusing on the missile for another ten seconds in case of any problems. (Image courtesy of the Airborne Laser Project.)

The adaptive large optics technology developed at the weapons lab also laid the groundwork for the Airborne Laser as it emerged in the 1990s. That program had a mission fundamentally different than the early Airborne Laser Laboratory. It was to fly at an elevation of forty thousand feet, where it would identify and target ballistic missiles as they rose out of the clouds a few thousand feet below, then zap them at distances to a few hundreds of kilometers. The air is thin at that altitude, but the mission required compensating for atmospheric effects along the entire path in order to deliver the lethal laser energy. Achieving that required aiming a laser beam from the aircraft at a vulnerable part of the target, detecting the reflected light, and measuring what happened on its round trip so the adaptive optics could compensate for the perturbations and deliver the high-powered beam to the target. Integrating those adaptive features into the optical system was difficult but essential for the Airborne Laser to succeed in shooting down missiles.[19]

Combined with advances in the short-wavelength COIL laser, that progress on adaptive optics convinced air force leaders that airborne lasers would offer a cheaper and more effective defense against ballistic missiles than space-based lasers could provide. In 1991, the air force started planning the Airborne Laser.[20] But it would take time to get the project going.

NARROWING THE FIELD OF LASERS

The first job was defining what they wanted to defend against. The Air Combat Command thought an Airborne Laser should target rapidly proliferating short- and medium-range tactical ballistic missiles during their vulnerable boost phase, from 150 kilometers (ninety-three miles) to several hundred kilometers away. That would drop dangerous debris back on or near the launch site and destroy decoys as well as multiple warheads when they were far more vulnerable to laser attack than above the atmosphere or returning to Earth. Those missiles make an easy target because they move relatively slowly, and the hot rocket exhaust makes them easy to spot. Past the boost phase, the laser would have to destroy submunitions that had already deployed, a more difficult task.[21]

The air force initially looked at chemical rocket-engine lasers as well as the COIL, and projected test flights of an airborne laser to occur before 2000.[22] But by the time it got down to competing conceptual designs, teams led by both Boeing and Rockwell International settled on COIL lasers in Boeing 747 planes. The two also faced competition from interceptors, which had their own attractions.[23]

After six months of lethality testing with MIRACL, lasers were looking good enough that the air force began talking about going beyond the original one planned test bed to a fleet of seven Airborne Lasers, with five always ready to go on patrol. The plan was for planes to fly figure-eight patterns near forward troops for up to eighteen hours, carrying enough chemicals to allow twenty to forty shots before refueling. "This is not smoke and mirrors," said Airborne Laser program director Colonel Richard D. Tebay.[24]

A NEW TYPE OF LASER

Killing boost-phase missiles at the desired range required a megawatt-class laser, and that required cranking up the power level of then current COIL lasers, which were rather different beasts than rocket-engine lasers. The chemistry is complex, requiring blowing chlorine gas across a solution of concentrated hydrogen peroxide to produce excited oxygen, and diluting that to mix with iodine molecules, forming the excited iodine atoms that emit laser light. A crucial advantage of that chemistry is that it can be run as a closed system that reuses many of the chemicals or converts them into solids so nothing has to be vented as exhaust. That's important because venting laser exhaust would push the plane, affecting beam direction.

Avoiding gas flow also prevents the vibration that made it almost impossible to point a rocket-engine laser steadily onto a target. "I had to design against vibration," recalls Ken Billman, who led the design of the optical system for the Airborne Laser. When TRW completed its COIL module, "you could put your hand on top of the chamber, and the only thing making any noise was the shower of chemicals being ultrasonically dispersed as it dropped down through the gain chamber by a small ultrasonic vibration." The internal operating temperature was only a couple hundred degrees Fahrenheit. "The whole thing was a quite benign yet powerful system," he says.[25]

That didn't mean chemical oxygen iodine lasers were completely fault-tolerant. Concentrated hydrogen peroxide is a hazardous material, particularly when mixed with strongly alkaline materials such as sodium hydroxide, which is done in COILs.[26] But that should be expected when working with the highly energetic chemicals needed to make chemical lasers.

AIR FORCE LASER PLANS

At the end of 1994, the air force estimated—guesstimated would be more accurate—that building and testing the first flight demonstrator would cost $600–$700 million. The plane would carry a megawatt-class COIL and fly at forty thousand to fifty thousand feet, ready to shoot at boosters after they emerged from the usual cloud deck at thirty-eight thousand feet. "There will be money battles," predicted Colonel Pat Garvey of the Air Combat Command but added that "the technology is here to make this happen."[27]

The final plan emerged in August 1995. It envisioned a $6 billion program to build a fleet of at least seven Airborne Lasers, each loaded with enough fuel to zap up to forty ballistic missiles in their first two minutes of flight. That was a big bet on new technology, but interceptors would have cost $45–$50 billion. Planners also envisioned two other roles, targeting low-flying satellites or enemy aircraft. However, at the time no laser kill of a boosting ballistic missile had been demonstrated.[28]

As the time for the contract decision neared, air force officials kept launching new ideas. One was surrounding the lumbering 747 carrying the laser with a pack of fighters in case its own laser couldn't fight off attackers. Another was to expand its range.[29]

A GOOD IDEA AT THE TIME

A dozen years after Reagan launched Star Wars, the Airborne Laser seemed somehow plausible. I recall the optimism of an engineer from the MIT Lincoln Laboratory giving a local talk. He admitted that orbiting Star Wars laser battle stations burning chemical fuels were science fiction. We didn't have the money or the technology to build them, or to zap nuclear missiles thousands of kilometers or miles away. And with the Cold War over, we didn't need those capabilities.

But he saw an Airborne Laser as within reach. A 747 can carry a hundred-ton payload in the air, four times what the shuttle could carry to low earth orbit at a vastly greater cost, and eight times what it could carry to the International Space Station.[30] Civilian 747s flew many times a day; the four then existing shuttles flew a few times a year.

Moreover, the Airborne Laser's goals were more modest. At worst, it only needed to zap a few "theater" ballistic missiles launched by a "rogue state" like North Korea or Iran. Those missiles could carry highly destructive warheads, but their range was limited to hundreds of kilometers or miles, far short of the distance to the US mainland. That seemed easy compared to Reagan's plans for a leak-free nuclear umbrella to stop thousands of Soviet nuclear missiles.

Yet looking back, I can see the same eagerness to believe that Gould saw in the colonels at ARPA. Ultimate weapons always work better on paper or in PowerPoint than in the real world.

DESIGN OF AN AIRBORNE LASER

The components of the Airborne Laser were tightly packed together in the 747 aircraft. The back two-thirds of the aircraft were filled with the laser itself and its fuel. The six SUV-sized laser modules were near the back, along with the chemicals and the processing system. They were designed to operate at ambient pressure at an elevation of forty thousand feet. Each laser module and the chemicals it contained weighed about 6,500 pounds. Two solid-state illuminating lasers also were housed in the back area. An airtight bulkhead separated the laser and chemicals from the rest of the aircraft to assure the safety of pilots and the operations crew. The laser exhausted water vapor and residual amounts of other chemicals through vents, not shown.

The battle management system, an active ranging system built around a carbon dioxide laser, and the beam control system were mounted in the front third of the main compartment. Pilots sat above the front part of that compartment.

A beam-delivery tube passed from the COIL laser in the rear of the plane through the front third and into the beam control system. The high-energy laser beam and the other laser beams were directed into an optical turret mounted on the nose of the plane, which could rotate and direct the beam as needed. The turret contained an adaptive-optic output mirror that would compensate for atmospheric effects. The beam emerged through a 1.5-meter window in the turret (see photo insert).

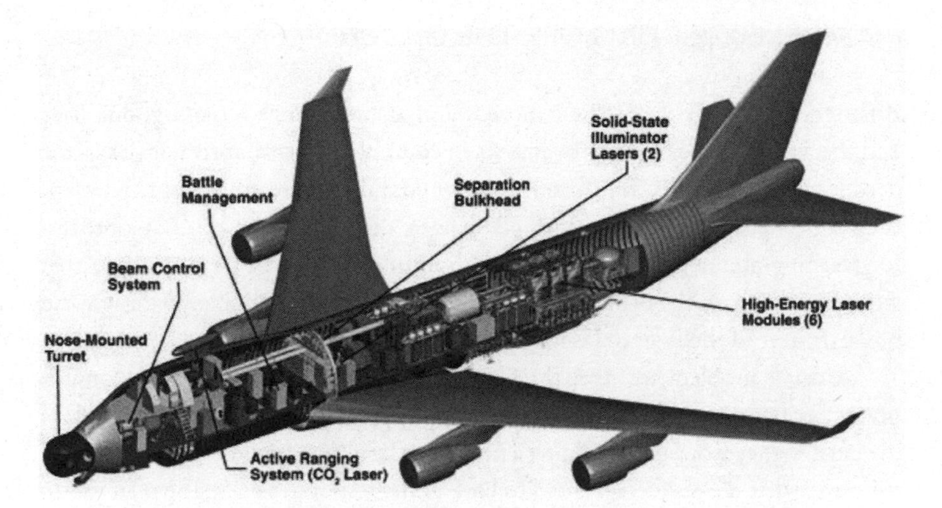

Figure 6.2. Inside the Airborne Laser. The chemical oxygen iodine laser filled the back two-thirds of the plane, with tubes delivering the laser beam through the battle management zone to the beam control system including adaptive optical system in the front. Beams from the solid-state illuminator lasers passed through the same tube. The turret on the nose of the plane turned to follow the target. (Image courtesy of Boeing.)

Three smaller lasers made up the pointing-and-tracking system. A carbon dioxide laser emitting at 10.6 micrometers tracked potential targets. A solid-state track illuminator laser emitting at 1.03 micrometers illuminated the target to examine it, locating its nose and the fuel tank, which is the weak spot targeted in liquid-fueled missiles; the heat from the COIL laser weakened the pressurized walls, which rupture when softened. (Solid boosters are less vulnerable, but heating can damage the outer wall.) The 1.06-micrometer beacon illuminator laser was focused on a spot on the target so the adaptive optical system could use the reflected light to compensate for atmospheric effects.

PROBLEMS GETTING OFF THE GROUND

Boeing's "Team ABL" got the contract, and at first things looked good. They had the advantage of big improvements in computers and adaptive optics in the decade since Star Wars. But in trying to build a fleet of seven Airborne Lasers, they faced many other technology challenges, so the air force built in a process for learning about new technology by building and testing a prototype that would be the first Airborne Laser and then using that knowledge to design and build the rest of the fleet of laser planes.[31]

A major problem was that the COIL was not yet able to generate as much power per pound as the air force wanted. The size of the 747 limited the bulk of the laser, so they wound up building a prototype laser that would only generate half the power that the air force wanted to have in the rest of the fleet. The plan was to try shooting down targets with the half-size laser, and to use what they learned from those tests to design and build full-power versions of the laser to put into the rest of the fleet of seven laser planes.[32] However, many things did not work as expected.

One was the chemical oxygen iodine laser itself. Although the air force called COIL technology mature, no one had measured the output power and the quality of the laser beam simultaneously. That's important because a laser weapon needs both a narrow beam and a high power to deliver the high intensity required to kill a target. Questions also arose about the August 1999 explosion at a test facility of a three-hundred-gallon tank filled with concentrated hydrogen peroxide, one of the chemicals used in the system.[33]

The Airborne Laser also ran into some unexpected problems in getting the optics it needed. With three separate lasers, each emitting a different color and needing its own beam control system, that meant a lot of devices. In fact, the prototype COIL being used in the initial Airborne Laser was being built as six separate modules, each about the size of a Chevrolet Suburban, and each with its own optics. The follow-on full-range capability second laser aircraft would require packing fourteen such gain modules into its 747.[34]

What the air force had thought would be a good period in which to build a new type of optical system turned out to be a bad one. The problem was not their technology but the dramatic success of fiber optics and the internet, recalls Ken Billman, who designed the beam control system being built by Lockheed. The spread of the World Wide Web had created a tremendous thirst for bandwidth that could only be supplied by fiber optics. That was creating a huge demand for optical components, and a huge demand for optics specialists was luring away all the people Billman needed to make those optical components.[35]

Whole companies were getting pulled into the economic boom. Itek, which had built a reputation for excellent optics and was a pioneer in adaptive optics, had bet big on Star Wars and built a massive coating chamber for large space optics such as needed by the Airborne Laser, only to see the Star Wars budget fall nearly as fast as it had risen. By the 1990s, Hughes had bought Itek and was selling off pieces of the company.[36] Suddenly the precision optical coatings Billman needed were no longer available.[37]

He tried Optical Coating Laboratory Inc. (OCLI) in California, which as its name says made optical coatings. They were interested in making the coatings for the Airborne Laser, but the telecommunications industry also wanted their filters to increase the capacity of their transmission systems. Communications buyers outbid the air force, and in 1999 OCLI itself was bought by a bigger optics company for $2.8 billion,[38] and all their coatings went into the commercial optics market.

Billman chased down a group of former Itek coating engineers who had joined a small Boston-area company called Barr Associates that specialized in optical coatings. There he found an owner with a patriotic bent who was willing to help. Barr already had two big evaporation chambers turning out coatings, but he had room for a third and would install one if Lockheed paid for it. "It wasn't really that much, but when I brought the offer back, Lockheed said they didn't have the money," Billman recalls. He says such coating issues caused two or three years of schedule delays.[39]

Other optical supply problems kept adding up. Long lead times of three years in delivering the laser turret and two years in delivering the 747 to house the laser left little time for careful testing, the Congressional Research Service warned. They also worried about testing issues and delays arising from the integration of subsystems.[40] Air force plans to restructure the program threatened a two-year delay in shooting down a missile.[41]

In short, what the air force had hoped would be a model fast-paced program was emerging as a textbook example of a troubled one, plagued by delays and cost overruns. Those problems threatened to cascade because of plans to start producing the next plane before the first one had finished testing.

A CHANGE IN COMMAND

Colonel Ellen Pawlikowski was named director of the Airborne Laser System Program Office at Kirtland Air Force Base in April 2000. She knew little about lasers, but it was a chemical laser and she was a PhD chemical engineer with experience in managing technical programs, so she got the job just as the program was going into the critical design review, which she described as "the final review before you actually really get into the serious business of bending metal."[42]

Pawlikowski got into engineering and the air force at the New Jersey Institute of Technology, then a commuter school based in Newark. She liked the atmosphere and camaraderie of the ROTC program, and after summer and part-time work at a private company, she decided she wanted to use her talents for greater purposes than just making money for somebody else. Her knack for thermodynamics led her to a doctoral program at Berkeley, and once she started in the air force, she liked the work, particularly the problem-solving side of engineering. The Airborne Laser would give her plenty of opportunity in that line.[43]

She found a solid group of laser experts at Kirtland, many of whom had worked on the Airborne Laser Laboratory. "The directed-energy business is still a fairly small community," she says. In fact, it also was a family business. Don Lamberson, the driving force behind the older system, was on the independent review team for the Airborne Laser as a retired lieutenant general. His son Steve was chief scientist for the newer system, and Pawlikowski found him to be a cornerstone of that base of expertise.[44]

One big challenge she faced was that the giant laser was technically very difficult and pushed the limits of design. "The design of the system required a very high speed turbo pump to get the laser cavity to the right pressure, and it had to fit into a very tight spot. The very first time they turned it on, just to run it, the blades inside the pump collided and we ended up with a bunch of shredded metal in the cavity of the pump. It took close to a year to redesign," she said. Another problem came from using a composite material in the laser chamber to try to reduce its weight. "The first time we pulled a vacuum on the first one we built, it collapsed on itself. We had to go back and make it out of titanium, and we ended up having to hire just about every thin-walled titanium welder that the US had to try [to] get them all built in time," she recalled. "I'm an engineer. The devil's always in the details for us engineers to sort out after the scientists figure out the cool science. If the big window that took five years to build got cracked or chipped, we might have to start all over again."[45]

Vibration was a big issue. "The structure of the airplane flexes, not to mention turbulence in the air, and most precision optics before this had been built on granite tables," she said. But they had to keep the vibration down low enough that the laser could stay focused on a small spot on a moving missile a hundred kilometers (sixty miles) away long enough to knock it out of the sky.[46]

Countless technical problems were solved along the way. All six laser modules were supposed to fire at the same time when big valves opened up simultaneously. Only they wouldn't all fire at the same time. "We scratched our heads, we did all the calculations, and there was absolutely nothing that would say why one would not open and the others would," said Pawlikowski. Then a TRW engineer thought the cause might be a small pressure difference, so they added some new plumbing and flowed helium to equalize the pressure. "As soon as we did that, it worked every time."[47]

The Airborne Laser Laboratory experience helped, but not all of the lessons learned in the 1970s worked in the twenty-first century. The Airborne Laser needed a hardwired abort system to shut down the laser if something was going wrong that might damage the plane or the laser system, but the system kept shutting itself down before they could test it. The answer was simple, the old-timers said. They just taped the safety system open when they did the first few tests. "In this day and age there was no way I could just tape everything open," said Pawlikowski.[48]

Other big challenges came with George W. Bush's December 2001 decision to build a missile defense system and withdraw the United States from the Anti-Ballistic Missile Treaty. The Airborne Laser had been conceived to defend against relatively short-range ballistic missiles used in situations such as the first Gulf War. Planners had decided that the most effective defense would be firing powerful laser beams from high-flying planes at the missiles as they rose above the cloud deck and into space. Those missiles were limited in range and armed with conventional weapons.

The United States had already been developing missiles to destroy strategic nuclear warheads flying above the atmosphere during the midcourse part of their flight. However, warheads above the atmosphere were tough targets to hit, and Star Wars studies had shown that the missiles boosting the warheads into space were easier to hit, and could also destroy decoys before they were deployed. The only technology the Pentagon had in its development pipeline that was designed to destroy missiles boosting out of the atmosphere was the Airborne Laser, but it was designed for use against short- to medium-range ballistic missiles used in regional conflicts. For that mission, the laser-laden lumbering 747 could get to relatively close range. The process of shooting at a boosting ICBM was similar, but any rogue state capable of building one would have more sophisticated weapons that could target the laser-armed plane at a longer distance from the launch site. To deliver lethal laser energy over that longer range, the plane would have to fly at a higher altitude where the air was thinner and would allow more of the emitted laser power to get through and destroy the target. That made defending against long-range launches by rogue states harder for the Airborne Laser than zapping theater missiles in a regional war.

Backing out of the ABM Treaty was controversial, so opponents to the move targeted the Airborne Laser budget for cuts to stop development of the nuclear missile defense system. The Airborne Laser program found itself with a budget in the $200 million a year range designed for its original modest task, only a small fraction of the billions allocated for the midcourse defense system, despite the fact that antimissile missiles had been demonstrated before but boost-phase shoot-downs had not. Pawlikowski made it her job to take on those larger-scale problems so the engineers and scientists could focus on doing their jobs.[49]

With all those challenges, it was no wonder the Airborne Laser fell behind schedule as they ran over budget. The overruns came, the General Accounting Office (GAO) wrote in May 2004, "because the program did not adequately plan for and could not fully anticipate the complexities involved in developing the system." The total had already passed $2 billion, and more increases were on the way.[50] The schedule was in similar disarray, and the GAO said the military utility of the half-power prototype of the first plane in the planned series of seven was "highly uncertain."[51]

FIRST LIGHT FROM A GIANT LASER

The Airborne Laser was complex and costly, but the air force and contractors were making progress, solving problems one by one. On November 10, 2004, they reached a crucial milestone, "first light," the first time the laser produced a powerful beam of light.

"One of the highest points in my career was when we got first light on that high-energy laser," said Pawlikowski in 2018. That's saying something when another high point for Pawlikowski was becoming only the third female four-star general in air force history in 2015.[52] She now commands some eighty thousand people in the Air Force Materiel Command and manages $60 billion a year in spending.[53]

Figure 6.3. General Ellen Pawlikowski. (Photo courtesy of the Department of Defense.)

It took four years of intense work by the whole team responsible for the laser. The laser was on the ground, not yet installed in the plane, so they could work on it easily. "We were always under huge pressure to get things done, and we were always behind schedule because the schedules were always way optimistic," Pawlikowski said. She was at an off-site meeting when her team called from Edwards Air Force Base in California to say they were ready to try the first test, but after all the troubles they had weathered, they were not optimistic. She took a red-eye flight back, and she turned on her cell phone when she landed. "I had a text and I had a voice message from Steve Lamberson, and all it said was 'we have a laser.' I'm like jumping up and down in the airplane, and people are looking at me like 'what the heck is wrong with you?' but after all these years, I was ecstatic," she said with a sigh.[54]

The team worried so much about damaging the laser that they only fired it for about a second, but it was a tremendous milestone. When Pawlikowski left the program in March 2005, the lab gave her the calorimeter used to measure the power of the laser beam in that first shot. "It's a big chunk of copper," she told me when I interviewed her on the phone. "I'm looking at it right now, and even now I can see the octagonal shape of that beam. Whenever I get discouraged at things around here, I look over at that plate and it reminds me that persistence pays off and sometimes you just have to keep your head down and just keep working through those 'just engineering' challenges."[55]

I was on the press teleconference in which she announced the results six days after the laser's first light. She said it verified that the design had the right physics to create and sustain a megawatt-class laser beam. She said engineers had spent eight days checking that all the subsystems ran properly. They planned further tests the following month, then a series of ground tests with higher and higher power until the laser can sustain full power for long enough to zap a missile. They also planned tests of the 747 modified to carry the laser, with a movable turret installed in its nose to direct the laser beam.

THE MOST COMPLEX WEAPON

In 2007, *Aviation Week* summed up the Airborne Laser as "the most complex integration ever attempted in a single weapon. There are 3 million lines of computer code to be integrated into a modified Boeing 747-400 transport aircraft with a total weight for modules, fuel, optics and subsystems of 50 tons, designed to fly at 40,000 feet."[56]

That said a lot in a few words. The people who worked on it understood the real complexity. "All expert technical observers agreed that the hardest Airborne Laser technical issues were beam control and fire control," said Ken Billman, who led the design of that part of the optical system. "How do you shoot this laser beam over this distance and keep it steady on a twenty-two-centimeter-wide target moving at kilometers per second? Every bit of beam jitter loses energy," and losing any energy makes it harder for the laser to do lethal damage in the limited time it has to deliver energy to the boosting target.[57]

The complexity of the optical system came from the need to do many things in the process of zapping missiles as they rose above the cloud deck. The original design included three smaller lasers to help find and track targets and to help deliver a lethal dose of high-energy laser power to destroy them. A small electrically powered carbon dioxide laser mounted on a pod on top of the plane was intended to initially spot targets. Its ten-micrometer wavelength could detect a clear path, free of any ice crystals at high altitudes where the Airborne Laser was designed to operate. The plane was to use its imaging system to pick out the optimum target for the laser to attack, but the pod air drag proved troublesome when the plane was flying, and the laser didn't work very well, so the pod and laser eventually were pulled off the plane.[58]

Two other cutting-edge solid-state lasers were inside the plane, operating at powers high enough to probe the target and return information signals to the Airborne Laser. The tracking illuminator laser was intended to track the target's course and identify where to direct the high-energy beam to do the most damage. The beacon illuminator laser then focused on that vulnerable "sweet spot" so sensors monitoring light from that laser reflected back to the plane could measure atmospheric effects on the path to the target, and the adaptive optics that aimed the high-powered COIL laser could be shaped to compensate for the measured atmospheric effects and focus the beam tightly.[59] To keep everything aligned, all the light from the tracker and beacon lasers, together with that of the COIL laser, was passed through the same set of optics, a "Common Mode, Common Path" technique that Billman said "has so many benefits that I would never have been able to design the Airborne Laser optical system without it."[60]

The light-generating COIL modules were near the back of the plane, with the optical alignment beam control system in the front, just behind a movable turret that had been installed at the nose of the plane to aim and focus the laser beam onto the target. With these vital components on opposite ends of the very lengthy plane, which was flexing along its length, this presented a major alignment control problem. Billman's group then developed a technique to automatically realign the path of the laser beam to correct for the continuous aircraft flexure.

"It's all done by magic," he said with a laugh after he had talked me through a series of diagrams. It made me think of Arthur C. Clarke's famous observation: "any sufficiently advanced technology is indistinguishable from magic."[61] It's my favorite Clarke quote because it's an elegant truth. Nikola Tesla would have understood mobile communications, but the electronics inside a smartphone would have seemed magical to him. To the average senator or representative or general, the elegant optics inside the Airborne Laser would have seemed magical. Yet being indistinguishable from magic was not enough to make the Airborne Laser an ultimate weapon.

After the COIL's first light in a ground test, the Airborne Laser team plowed onward. They test-flew the 747 modified to carry the big laser. Then they installed the optical systems, auxiliary lasers, and a lower-power test laser to assess key technical issues in controlling the powerful laser beam when it was installed, perfecting the adaptive optics needed to control atmospheric perturbations, controlling vibrations and their impact on the optics, and integrating all the components into a working system. It was a meticulous step-by-step process of debugging, looking for things that didn't work quite right, then fixing them and making sure the fix didn't break anything else.

Weight had long been a crucial issue because it's always a limiting factor in aircraft performance. Initial plans had called for installing fourteen COIL modules to provide megawatt-class power, but the COIL modules proved larger than expected, with six weighing in at a total of 180,000 pounds, heftier than had been planned for the full complement of fourteen. The Airborne Laser team planned the first plane as a demonstration version to work out engineering details, which did not have to deliver full design power, so they installed only those six modules. They planned to shrink and redesign the modules so they could install enough of them to reach full design power in a second plane, which was to be an engineering/manufacturing demonstration. That was to be the basis for producing a full fleet of seven weapon-grade Airborne Lasers with one that could be serviced as three were used in each of two simultaneous military actions.[62]

OPERATIONAL QUESTIONS

In 2007, it was becoming clear that the biggest problem with the Airborne Laser might be "CONOPS," military jargon for the "concept of operations" or how it could be used in military operations. "As the [Airborne Laser] program nears procurement and potential fielding, questions remain about the number of aircraft to be procured, where the aircraft might be deployed, and how they would be used," the Congressional Research Service (CRS) wrote in a review paper. In other words, if this thing does work, how can the military use it effectively?[63]

The size of the aircraft and its need for special materials and aircrew created problems. The planes would have to be based "forward," close to the action so they could strike quickly when needed. But their need for special chemicals that required sensitive handling and equipment not available on the front lines would mean that once an Airborne Laser had burned up all its laser fuels, it would have to go back home for recharging. That could open an opportunity for the enemy to attack while the crucial weapon was back home in the shop.[64]

The Pentagon had planned a fleet of seven Airborne Lasers, but the CRS warned that with the need to refuel the planes, replace expended fuel, and handle maintenance and crew down time, a fleet of seven might serve only one of the air force's ten separate expeditionary wings. Would they be split among the forward groups, or all of them concentrated in a single wing?[65] The first would mean each group had an Airborne Laser available only a small fraction of the total time, so an enemy could run out the clock on the laser, then come in for the attack when the laser had to fly home after using up its time or fuel.

The second would limit defense to only one forward group, inviting an enemy to target other forces left vulnerable because they lacked lasers. With hot spots spread around the world, modern armies need to redeploy quickly, and that would be hard with so few lasers.

The laser would be "a highly visible target," massive, lumbering, and requiring an escort of fighter jets for protection. The navy had already seen that problem with its massive battleships vulnerable to small boat attacks by insurgents. The air force and other services had already complained about the heavy costs of providing operations and support for other "high demand, low-density assets" like the sophisticated 707-based JSTARS airborne battle management system that was valuable but few, far between, and vulnerable. A fleet of Airborne Lasers also could require more refueling from the limited fleet of aircraft dedicated to in-flight refueling.[66]

The mass of the laser, its fuel, and associated equipment posed another logistic complication. The second and later Airborne Lasers were intended to have more modules and higher power that would be attractive because it could take on more targets in a shorter time. But the prototype aircraft came in overweight despite only including six laser modules rather than the planned fourteen needed to extend the laser's target range and effectiveness. The more laser modules, the more fuel that would be needed, and the more aerial refueling would be needed, a general drain on resources. The lower the fuel levels on board, the more often the laser planes would have to fly back home to fill their chemical tanks. As long as the range remained short, the lasers couldn't fly over Russia and China, but that might not be a good idea anyway, because the Airborne Laser would be vulnerable to antiaircraft weapons.[67] The agency also complained, "It is difficult to assess how the ABL [Airborne Laser] might be employed because it is not currently clear what the ABL's capabilities will be, once fielded."[68]

Yet with the first target shots seeming to be just around the corner, give or take a few hundred million dollars, neither Congress nor the Bush administration were ready to pull the plug on what many still hoped would be a hot technology, if not the ultimate weapon. There also was hope that better lasers were on the way. Indeed, the power of the long-neglected solid-state lasers had been rising steadily in industry, and military labs had been reporting encouraging news, covered in chapter 8.

FIRST TARGET SHOTS AFTER $4.3 BILLION

The Airborne Laser, designated YAL-1A, fired its first shots at targets on November 24, 2008, while sitting on the ground at Edwards Air Force Base. It was a milestone, four years after first light. The laser fired for one second at each of two different targets. A dozen years had passed since the program started in 1996, when it had plans for first target shots by 2002. By then its budget had bloated to $4.3 billion, a factor of four above original plans.[69]

With plans for the first lethal tests in 2009, Boeing kept a brave face. "There's nothing like flaming missile wreckage to show the world the system is viable and it works," said Mike Rinn, Airborne Laser program director at Boeing. He said the big laser was optimized for the ballistic missile defense the Pentagon was pushing, but could be adapted for other targets, including aircraft, surface-to-air missiles, and cruise missiles. There had been talk of adding satellites to the target list, either by blinding spy cameras or destroying the tanks containing the fuel for maneuvering the satellites. But the laser had yet to run on a plane in the air.[70]

The pace of tests continued slow and deliberate. When the Obama administration took over, carry-over defense secretary Robert Gates took a fresh look at the budget. Despite the delays and cost overruns, the Airborne Laser had officially remained a program intended to deploy an operational weapon system. Gates said it should be slipped back to research and development and a proposed second plane should be canceled because its proposed operational role was "highly questionable" and faced cost and technology problems.[71]

"I don't know anybody at the Department of Defense . . . who thinks that [the Airborne Laser] should or would ever be operationally deployed," Gates told a March 19, 2009, hearing of the House Appropriations Committee's Defense Subcommittee. "The reality is that you would need a laser something like 20 to 30 times more powerful than the chemical laser in the plane right now to be able to get any distance from the launch site to fire."[72]

"So, right now the ABL would have to orbit inside the borders of Iran in order to be able to try and use its laser to shoot down that missile in the boost phase. And if you were to operationalize this you would be looking at ten to twenty 747s, at a billion and a half dollars apiece, and $100 million a year to operate. And there's nobody in uniform that I know who believes that this is a workable concept."[73]

Gates's words weren't enough to kill such a big program with testing already in progress. Its 2009 budget was already allocated, and the budget proposed for fiscal 2010 was $187 million. The air force still could learn from tests, and officially the Airborne Laser became the Airborne Laser Test Bed. In June tests, Boeing said the Airborne Laser detected missiles emerging from the cloud deck at a "militarily useful range." The Track Illumination Laser locked onto the target point on the nose of the missile, and then the Beam Illuminator Laser locked onto the target, creating a control loop to compensate for atmospheric turbulence and jitter. Finally, a third laser emitting at the same wavelength as the big chemical oxygen iodine laser, but with much less energy, illuminated the target with too little power to damage it.[74]

Not until August did the Airborne Laser fire its massive weapon laser in the air, but the beam went into measurement instruments. In February 2010 the big laser intercepted, but did not destroy, a test target while in the air. Next it destroyed a liquid-fueled ballistic missile and engaged a solid booster but did not destroy it. But officials warned of flaws in beam alignment that needed to be corrected for the laser to function properly. The test flights cost about $30 million. Some observers were encouraged by the test results, but Gates and others remained doubtful.[75]

Gates's blunt assessment reflected operational reality. COIL technology was the best to date, but it still limited what a fleet of Airborne Lasers could accomplish. Each plane could carry only a limited amount of laser fuels and would have to fly far from the conflict zone to replenish the fuel because the chemicals required special handling not practical in the field. The planes would have to keep flying around to maintain their stations, which could require expensive aerial refueling. Planned follow-ons promised more capabilities, but it was far from clear if those promises could be realized. And questions remained about what capabilities the Airborne Laser would have in the field.[76]

The handwriting was on the wall, but it took time to shut down the Airborne Laser. It still had advocates, but it had become a large and tempting target after running up huge bills with little prospect for usable follow-on development. Once the Missile Defense Agency had the big laser, they wanted to test its capabilities. But in November 2011, the Missile Defense Agency said it was done with flight tests and told contractors to remove the laser, classified hardware, and other materials from the plane. *Aviation Week* called it "a symbolic end to Boeing's preeminence as the leader on the Pentagon's once linchpin directed-energy program and its ambitions application as a missile defense system."[77] However, that was an overstatement. What it marked was the end of an era for big chemically fueled lasers because of their serious operational drawbacks. The Pentagon was already investigating other lasers.

LOOKING BACK AT THE AIRBORNE LASER

Specialists proud of their accomplishments were sad to see the Airborne Laser scrapped. Some remain angry. "I'm still pissed," said one developer involved since day one. He blamed its fate on shifting its mission from tactical use in regional conflicts to filling a role, in which it fit poorly, in a national missile defense system. "ISIS would never have been a problem if you had that thing running," he told me as I was writing this book.

Yet others recognize that it had serious limitations. The Airborne Laser was "an engineering tour de force" to Ed Gerry, who retired from Boeing in 2007. "The fact that it worked at all was an engineering marvel, and it was at least successful in doing that." He didn't see the sense in trying to base an operational military system on a 747. Its range was limited, and to be useful as a missile defense system, the laser would have had to be much brighter.[78]

At the end of the program, Janet Fender says, "We had beautiful beam quality from the chemical laser."[79]

"We learned a lot about the engineering of these systems" from the Airborne Laser, said General Pawlikowski. "We learned an awful lot about pointing and tracking, which will be absolutely essential to any laser that we're able to put onto a fighter or any self-defense system. We learned a lot about atmospheric compensation, and some of that is being used in telescopes today. We learned solid-state lasers and their limitations." Working with a COIL helped us "better understand the logistical challenges of dealing with that type of laser in a battlefield environment." Most important, she says, "we have a whole generation of scientists and engineers and program managers that learned huge lessons from that experience, me in particular."[80]

With flight tests over, the Airborne Laser was flown to the air force's "boneyard" for long-term storage at Davis-Monthan Air Force Base in Tucson, Arizona, in February 2012.[81] Officially known as the 309th Aerospace Maintenance and Regeneration Group, it's the air force version of an auto junkyard, with the desert site chosen to preserve the old planes as long as possible. Salvageable parts are removed and stored, piece by piece, to be donated to keep other aging air force aircraft flying. Eventually the rusting hulks that remain are sold for scrap value.

The air force has a museum on the site, and Allen Nogee, a market analyst specializing in lasers and optics, took a bus tour of the boneyard in the spring of 2014. He saw the decommissioned Airborne Laser sitting by itself, "a $5 billion failed U.S. military experiment in directed energy weapons." In a matter of months all the pieces would be gone. Musing on how the civilian and military use of lasers has progressed in the twenty years since the Airborne Laser program was started, he wrote, "It's still a bit sad to see this one bit of laser history disappear."[82]

The old 747 may have been the sort of memory that top Department of Defense managers wanted to forget. But the Pentagon has not given up on all lasers. Chapters 8 and 9 tell how new programs have shifted to solid-state lasers and are exploring other types of lasers powered by electricity rather than chemicals. The hope is that they can tackle challenges at lower powers, and perhaps later be scaled up to the megawatt class needed for missions such as boost-phase defense with robotic drones rather than jumbo jets.

CHAPTER 7

BACK TO THE BATTLEFIELD AGAINST INSURGENTS

The ultimate weapon that Israel wanted in the mid-1990s was something that could stop the Katyusha rockets fired by Hezbollah insurgents from just inside the southern border of Lebanon. Deadly, cheap, and readily available, they rained death down on children and other civilians that no existing defense systems could stop effectively. With the United States gearing up to start development of the Airborne Laser, Israeli officials wondered if speed-of-light high-energy laser defenses could zap Katyushas before they could strike. It might not be an ultimate weapon on the global scale, but it would be one for Israelis who lived within rocket range of the border.

The world had changed after the collapse of the Soviet Union. Interest in space-based lasers as the ultimate weapon to end the balance of nuclear terror waned as the Cold War thawed. Insurgents and rogue states emerged as the main threats to national security, and the quest for the ultimate weapon turned toward lasers that could zap insurgent weapons into oblivion. After a long eclipse, laser weapon development was turning back to short-range weapons for use on the battlefield.

In 1995, the Israel Ministry of Defense and the US Army Space and Missile Defense Command formally launched a feasibility study of prospects for laser defense against insurgent weapons. It began with a series of lethality tests using the venerable MIRACL laser, the centerpiece of the US Army's High-Energy Laser Systems Test Facility at the White Sands Missile Range in New Mexico. The results were encouraging and climaxed in February with the shoot-down of a short-range artillery rocket with MIRACL attached to its Sea Lite Beam Director on February 9, 1996.[1]

MIRACL was far too massive to be usable as a laser weapon. What the Israelis wanted was a laser that was compact and transportable so they could haul it to sites where insurgents were operating and blast their rockets out of the sky. They and the Army Space and Missile Defense Command decided to take the next step and build a compact and transportable version of the deuterium fluoride rocket-engine laser design that could destroy insurgent rockets in flight.[2]

The plan was part of the Advanced Concept Technology Demonstration program established in 1994 to speed development of new weapon systems, which had become so slow that long years of development wound up costing buckets of money and delivering technology that was obsolete by the time it reached the battlefield.[3] Initially, the Tactical High Energy Laser (THEL) received only minimal funding from Congress, but it got a boost from the successful MIRACL tests. In April 1996 it got another boost from an unwanted source: Hezbollah militants who fired more than two dozen Katyusha rockets from southern Lebanon toward Israeli border towns in a two-week period.[4] That got attention from US president Bill Clinton and Israeli prime minister Shimon Peres, who agreed that the United States would help Israel develop a laser defense system. In July 1996, the United States and Israel formalized their agreement, and later that month the US Army awarded an $89 million contract to TRW, which had produced the MIRACL and ALPHA rocket-engine lasers, to build a prototype system. It included a deuterium fluoride laser emitting a few hundred kilowatts, a beam pointer-tracker system, and a command control and communications system.[5] The allies dubbed it the Tactical High Energy Laser Demonstrator (THEL Demonstrator) or the Nautilus Laser.

The promise of shooting down insurgent rockets was a strong motivation to complete the proposed laser system in short order. That meant using existing or readily improved hardware wherever possible, making interfaces simple, simulating as much as possible, and doing as many things simultaneously as they could. That required using the well-developed but inefficient 3.8-micrometer deuterium fluoride rocket-engine laser and optics, but the developers said the strategy worked very well. TRW was able to build the bulk of the laser hardware—the gain generator and the optics—within nine months after the design was finalized. It took another year to assemble the whole system and complete the rest of the components, and on June 26, 1999, they fired the laser for the first time.[6] More time was needed to move equipment from the TRW site in San Juan Capistrano, California, to the high-energy laser test facility at White Sands.

THEL shot down its first rocket in the air on June 6, 2000, just four years after the program started. "This would be remarkable enough for any weapon demonstration program, but was even more so for this fundamentally new type of weapon," wrote three developers.[7] More shoot-downs followed as the developers explored the range of potential conditions under which the laser would have to shoot down rockets. Those early tests shot down more than twenty-five low-cost, portable, and deadly Katyushas. They are hard to shoot down with conventional weapons, but they are relatively easy targets for lasers because they are pressurized, so a laser beam pointed steadily at the surface can heat the casing, softening it beyond its pressure tolerance so the pressure explodes the rocket and destroys it in the air.[8]

CLOSER TARGETS ARE EASIER TARGETS

At first glance the THEL success stands in remarkable contrast to the long delays and multibillion-dollar expense of the Airborne Laser. But to be fair, THEL was a much simpler project. The Airborne Laser was an inherently ambitious program, aiming at scaling the new COIL type of laser to megawatt-class output, cramming it all into an airplane, and delivering that megawatt-class beam through over a hundred kilometers of air as the plane flew at forty thousand feet. THEL needed a four-hundred-kilowatt-class rocket-engine laser, a power level that had been attained before, so it could draw on previously developed technology.

THEL was shooting at targets at comparatively near distances, which were not disclosed but probably were at most a few kilometers away. That's a big advantage because atmospheric effects that perturb or absorb build up with distance. A good quality beam from a laser weapon might maintain its high quality well enough to be lethal for targets a few kilometers away, especially if the targets were relatively soft. That would avoid the need for costly and complex adaptive optics to clean up the laser beam enough to kill such targets. The Airborne Laser had to strike targets through a hundred kilometers or more of air, so it needed adaptive optics to deliver a lethal dose of laser energy.

THEL was designed to destroy insurgent weapons, which are loaded with high explosives as well as rocket fuel. Artillery and mortar shells also carry explosives, so laser heat also can detonate them. Likewise, laser beams can ignite the engines on small boats and drones, starting with fuel fumes seeping from the engine. The Airborne Laser was built to shoot at ballistic missiles, which generally are sturdier, and its mission was later expanded to include boost-phase nuclear weapons, which are even better protected against heat.

Another important difference was that THEL was not so much a weapon system as an array of trailers with a beam director sitting on top of one. Some housed the laser subsystem, including the gain generator, the supply of rocket fuels, and the pressure recovery system. The beam director and pointer-tracker subsystem were separate. The command, control, and communications facilities were housed in a trailer separated from those containing the laser; it's always a reasonable precaution to keep people a safe distance from chemical fuels during operation. The radar needed for identifying and tracking targets also was separate.

In contrast, a big challenge with the Airborne Laser was fitting all the equipment into the airplane and getting it to operate while the plane was flying at altitude. A plane in flight vibrates constantly. A parked trailer stays still as long as it is properly parked.

The bottom line was that shooting down comparatively short-range rockets from the ground was a much easier job than taking on ballistic missiles from a flying airplane. Rocket attacks were also much more immediate problems than anything involving strategic defense. Israeli forces faced insurgents next door; over the next hill on the Lebanese border were Hezbollah fighters who could, if irked, start shooting rockets and killing people. Strategic weapons are shows of strength that do their job properly if they never have to be used.

The success of THEL would renew interest in bringing lasers to the battlefield in the new world where the Pentagon's big worry was warfare with insurgents. It was an idea that looked good in theory.

MOBILE THEL

THEL was a test bed used at White Sands. In principle, it could have been packed up and shipped to Israel for installation there, but at best it was "transportable" rather than mobile. Transportable means it can be moved but the process takes time and trouble, like removing a mobile home from everything it's attached to in one mobile home park, towing it to another site, and installing all the utilities. Israel needed a laser weapon that was truly mobile, ready to roll to another site where it was needed in no short order, like a motor home or camper sitting in a driveway. So the United States and Israel agreed on a mobile THEL (MTHEL) as a follow-up system.[9]

The goal was to redesign THEL so it would fit into a few mobile vehicles that could be hauled in standard C-130 cargo planes if necessary. The laser could be packed into three trailer loads, including the laser, gain medium, and fuel tanks for the first system. The long-term goal was to shrink it down to the size of a Humvee. The fire control radars and laser fuel would be on separate vehicles. Several MTHELs could be built, and they would be mobile enough to move to wherever they were needed to stop missile attacks. The details were to be arranged by a joint effort of the US Army and Israel, with the United States paying the bulk of the money.[10]

Some Israeli officers were worried that MTHEL would be unable to cope with the relatively recent development of Katyusha rockets with ranges to the tens of kilometers, which could produce much greater damage. "THEL can't give a full answer to the [Katyusha] missile threat even if we complete development and purchase huge numbers of systems. Although the technology is promising and can serve the industrial interests of both of our countries, this is not the way to cope with the threat of rockets fired from Lebanon," said Israeli major general Giora Eiland, head of the planning directorate of the Israeli Defense Force. Instead of building a fleet of MTHEL vehicles, he said, the Israeli military would rely on a massive conventional response against Syrian interests in Lebanon.[11]

Work started with a contract issued to TRW on June 12, 2001, to perform a trade-off study and analysis. In late 2002, the US Army tested the laser against artillery shells, which are only about two feet long, compared to the ten feet of a Katyusha, and which generate less heat than the rockets, making them harder to track. Within seconds, the energy delivered by the laser destroyed the shell, well short of its target.[12]

THE ADVANCED TACTICAL LASER

A second spin-off of THEL was the air force's development of the Advanced Tactical Laser (ATL) for battlefield use. The initial plan called for installing a small chemical oxygen iodine laser emitting fifty to seventy kilowatts in a tactical aircraft, where it could serve as a "laser gunship" to ignite fires, target vehicles, and scare enemy soldiers.[13]

The air force approved a $176 million contract to Boeing in 2002 to explore the use of a one-hundred-kilowatt-class COIL to be used from the air against "tactically relevant" targets on the ground. To carry it, they chose a C-130 Hercules, a venerable four-engine turboprop plane that the air force has used since the 1950s to carry troops and cargo and to wield weapons including Gatling guns, howitzers, and bombs. The C-130 has a nominal cargo capacity of twenty-two tons and a cargo space about one-sixth the size of that in the more cumbersome 747 that carried the Airborne Laser.[14]

A key attraction of the C-130 was the maneuverability needed for active air-ground combat. Neither the 707 used in the Airborne Laser Laboratory nor the 747 of the Airborne Laser had that agility; they were transport planes that needed airports and runways. The COIL to be installed on the C-130 laser gunboat would be the first high-energy laser designed to be fired from the air at ground targets. Air force planners hoped that the laser would live up to its promise as a directed-energy weapon by destroying or disabling military targets in urban combat zones without the collateral damage inevitable from conventional weapons. Insurgents were prime targets, and US and Israeli officials hoped the laser could knock out rocket launchers that insurgents were operating from civilian sites such as schools or hospital courtyards.[15]

The power level was unofficially reported as in the one-hundred-kilowatt class, but it was not just a downsized version of the megawatt-class Airborne Laser. The ATL laser module that occupied much of the smaller plane's cargo space was operated at cryogenic temperatures. It also was a closed system that pumped all the exhaust gas into a spent-gas tank, a design that lets it operate at low altitude, where air pressure is high and the gas might not flow as well as at low pressures and high elevations, and where releasing the gas could endanger personnel. In contrast, the Airborne Laser was designed to operate at forty thousand feet, allowing it to vent water vapor and some traces of other gases into the atmosphere at the low prevailing pressure. Another important difference is that the tactical laser did not require the complex adaptive optical system used in the Airborne Laser because its targets were within a few kilometers, not tens or hundreds of kilometers away.[16]

A COIL is a more manageable beast than the THEL hydrogen fluoride laser, but a one-hundred-kilowatt version was still a big thing to put into a relatively small cargo plane. Boeing spent more than a year working on the wingless fuselage of a retired C-130 to figure out where to put the laser, optics, controls, beam director, sensors, and other equipment. Then they installed the laser, control system, and other equipment in the plane that would fly with the laser inside it.[17]

The most obvious external feature setting the ATL apart from other C-130s was a rotating turret below the fuselage, which directed the laser toward targets on the ground. The geometry did not let it fire at other planes or targets in the air.[18]

Boeing started ground tests in early 2008 to identify problems and fix them before starting flights. That took a while. The laser wasn't able to hit a target board on the ground while flying in the air until June 13, 2009. On August 30, its battle management system focused the laser beam long enough on an unoccupied truck parked on the ground for Boeing to say the laser had "defeated" it and claim success for the ATL.[19] The moment of burn-through can be seen from two viewpoints on YouTube.[20] On September 19, the laser succeeded at the tougher task of burning a hole through the fender of a remotely controlled truck moving on the ground as the laser plane flew overhead. That was good enough for Gary Fitzmire, director of the Boeing Directed Energy Systems program, to hail "the transformational versatility of this speed-of-light, ultra-precision engagement capability that will dramatically reduce collateral damage."[21] The power the laser attained was not announced, but unofficial sources say it was tens of kilowatts.

Operations specialists were less impressed but were willing to consider a laser weapon as long as they could continue arming C-130s with one of the two big guns normally used on the plane. They weren't willing to part with the howitzer that fired 105-millimeter shells powerful enough to take out a tank. But they said they would consider removing a thirty-millimeter (Gatling) gun that was still a formidable weapon. They might have been kidding. In reality, there was no chance of squeezing the ATL that had filled the inside of the plane into the much smaller space occupied by the Gatling gun.[22]

The Air Force Scientific Advisory Board reviewed the results and bluntly concluded, "There is no operational utility of the NC-130 Advanced Tactical Laser Advanced Concept Technology Demonstrator (ATL ACTD), but ... there are some key measurements that could be obtained from this platform to inform future laser gunship development efforts."[23] In other words, the laser may have bagged a few sitting ducks on the target range, but it wasn't ready for any real-world battles.[24] They found some long-term hope for the concept of a laser-augmented gunship and suggested the air force should add it to its future weapon technology plans. But they strongly recommended that the air force work on maturing solid-state lasers and focus its technology investments on developing a field-ready laser gunship.[25]

PULLING THE PLUG ON MTHEL

MTHEL didn't last as long as the air force laser gunship. The project continued through 2005, but by that time military systems engineers grew worried about the chemistry of the laser. One problem was supplying the laser with the chemical fuels it needed for operation. A second was the perils of using equipment that used toxic and corrosive fuels, and had dangerous exhaust gases, in the battlefield. They diverted the money that had been intended for MTHEL to development of new solid-state lasers that they thought would be better for attacking rockets, artillery, and mortars.[26]

That decision must have involved some interesting conversations. The laser developers thought they had found a solution to the tough problem of shooting down rockets and artillery shells launched by insurgents. Top officials saw the mobile laser rocket-zapper as a vital program. But field operations experts knew that performance of military personnel and equipment in the field depends on the supplies delivered to them. So they would have asked what the lasers needed.

The answer they wanted to hear was diesel fuel. Modern armies live on the stuff. Diesel powers the vehicles, and it fuels the generators that turn out the electricity that other equipment needs. It's flammable but not as incendiary as gasoline.

The answer they got was that the laser needed two other chemical fuels. It was bad enough that they were extra things to worry about. Even worse was that the fuels were both dangerous and corrosive and that the exhaust was hydrofluoric acid, so corrosive that it destroys the cornea, causing blindness. Doubtless, laser developers assured system operations teams that all the exhaust would be scrubbed and go into tanks. But field operations experts knew that other things could happen on the battlefield; one direct enemy hit on the exhaust canister and their troops could have a chemical weapon in their midst.

The system operations experts must have said something to the general effect of "NO!" Lasers might be a great idea, but they wanted something that fit with their logistics requirements. They knew the problems with complex logistics. Weapons that need special fuel, or run out of ammunition or spare parts, end up dumped along the side of the road. They had enough dangerous stuff to worry about already. They insisted on laser weapons that could run on electric power from diesel generators, technology they already had on the battlefield.

The decision probably went up the chain of command, and the field operations side won, with good reason. It was time to look for alternatives.

LOOKING AT LASER ALTERNATIVES

The air force panel also made another important observation. "[T]he Air Force [should] mature solid-state laser technologies, and pursue the related system improvements in beam control, light-weighting, power sourcing, and thermal management."[27]

The air force had given the panel a broad charter, not just to look at the ATL but also to look at the whole current state of laser technology that could be used in laser-armed battle planes and their potential. That meant how laser weapons could be used for offense and defense, and how they might affect logistics, potential countermeasures and vulnerabilities, and future trends in the near, medium, and far term.

One crucial difference between tactical weapons and strategic weapons is that tactical weapons are made to be usable, but strategic weapons are made to look impressive. The goal of strategy is to convince the other side that starting a war is not a good idea. The goal of tactics is to win a battle. And the air force scientists on the panel couldn't see how the ATL could be much help.

The big rocket-engine lasers in THEL and MTHEL and the big chemical oxygen iodine lasers in the Airborne Laser and the ATL all had the same fundamental problem: they were too big, complex, and cumbersome to make effective weapons. They did convert more of the energy they generated into light than the Airborne Laser Laboratory had done two decades earlier. Back in its day around 1980, I recall hearing laser scientists joke that such lasers were so big that their only value to the military would be to drop them on the enemy. Hydrogen fluoride rocket-engine lasers had shorter wavelengths and were more efficient than gas-dynamic lasers. COIL lasers were a further improvement. But they hadn't improved enough.

In recommending solid-state lasers, the air force panel was saying the same thing that logistics officers had said when they took a hard look at THEL. Their message was that if we're going to use lasers on the battlefield, we must have better lasers. Fortunately, by the 2000s, those better lasers had come along in the form of a new generation of solid-state lasers. They had emerged gradually, largely from civilian efforts with assistance from military research projects.

THE SOLID-STATE LASER REVOLUTION

Solid-state lasers made from crystals, glass, and semiconductors got off to an early start in the 1960s. For a few years they produced higher powers than gas lasers and were the first incredible lasers the Pentagon investigated as possible death rays. Their fatal flaw was that they turned only a small fraction of the energy they absorbed into laser light, so they got hot and stayed hot for a long time, like a glass or ceramic baking dish just out of the oven.

The rocket-engine laser solved that problem by the simple expedient of blowing the hot gas out of the laser, taking the waste heat with it. Laser developers called it the "garbage disposal" principle, and it worked well in its time. Rocket-engine lasers made great demonstrations in the laboratory, but they were big and complex and not very efficient. But eventually their need for special chemical fuels proved to be a fatal flaw in plans to use them on the battlefield.

Fortunately, solid-state lasers had made tremendous advances since the 1960s. The first big improvements came in semiconductor diode lasers and opened the door to advances in other solid-state lasers. Military research funding was behind some improvements, but private funding led to others.

The first semiconductor lasers emitted only a feeble light and didn't last long. By 1977, Bell Labs had perfected the quality of semiconductor lasers so they could operate continuously for a century at milliwatt-level powers. Bell had developed the lasers for use in fiber-optic communications, but anyone could use the same technology for other purposes.

Military research programs sought to increase semiconductor laser power and efficiency by developing better semiconductor materials and internal structures to increase output power and improve how efficiently they converted input electric power into light. Commercial research sought new wavelengths for new applications that were not practical with the narrow range of invisible infrared wavelengths emitted by early semiconductor lasers. One dramatic breakthrough in the 1990s was the first bright blue and violet semiconductor lasers. Initially they were sought to record high-definition video on optical disks the same size as audio compact disks. Later, they gave rise to the blue LEDs that are the real color behind solid-state lighting (which uses phosphors to add the yellow that makes them look white to the eye). That invention earned three pioneers in the field the 2014 Nobel Prize in Physics.[28]

By the late 1980s, a new technology had emerged, using light from semiconductor lasers to power solid-state lasers. Strange as it may seem, using light from one laser to make another laser emit light makes perfect sense. Semiconductor lasers can turn over two-thirds of the electric current passed through them into laser light, vastly more efficient than the flashlamps that Theodore Maiman and Elias Snitzer used to power their crystalline and glass lasers in the early 1960s. However, semiconductor lasers do not produce the narrow beams needed for laser weapons.

Fortunately, semiconductor lasers emit only a single color of light like other lasers, and that turns out to be another big advantage for powering glass and crystalline lasers. The light-emitting atoms in glass and crystal lasers absorb light most efficiently at narrow bands of wavelengths. That means if you can make a semiconductor laser that emits exactly that wavelength, the light-emitting atoms in the crystal will absorb it all, and turn much of it into a tightly focused laser beam. If, instead, you illuminate the light-emitting elements with the white light from a flashlamp, only a small fraction of the light that happens to be at the absorbed wavelength will be absorbed, so the tightly focused laser beam will contain only a small fraction of the input light energy.

Nobody could do that in the 1960s, because semiconductor lasers emitted too little light for optical pumping and lamps had to be used instead. However by the late 1980s semiconductor lasers had become bright enough to power glass and crystalline lasers, and their power kept increasing in the 1990s. The semiconductor lasers were small, and each one was not very powerful, but the light from many of them could be combined to power a glass or crystalline laser that delivered much more energy than was possible in the 1960s. And it produced much less waste heat, so people didn't have to hide their heads under a metal trash can to protect themselves from overheated glass exploding, as Bill Shiner and Snitzer did in the 1960s.

The advance of fiber-optic communications also led scientists and engineers to stretch glass laser rods into long, thin optical fibers that acted as lasers. Directing light from a semiconductor laser into the end of an optical fiber proved to be a very good way to transfer power from the semiconductor laser into the fiber laser, and made fiber lasers very efficient. Some types were used to amplify light in fiber-optic communications networks. Others were used in industry because they could generate high-quality beams efficiently.

The development of high-powered solid-state lasers powered by semiconductor light sources that converted 20 percent or more of the input power to the semiconductor into a laser beam led to interest in using them as high-energy laser weapons. The most efficient types included fiber lasers, lasers in which the light-emitting solid was a thin disk, lasers in which the light-emitting solid was a thin slab, and all would be considered for weapon applications. It added up to an amazing comeback for a technology written off in the mid-1960s as impractical.

THE NAVY FREE-ELECTRON LASER

Some laser technology abandoned by Star Wars also was getting a second look from the US Navy after an October 12, 2000, small boat attack on the USS *Cole* killed seventeen sailors. Looking for a laser weapon usable at sea that could blow insurgents out of the water at short range, they turned to the free-electron laser, which Star Wars had abandoned as a long-range weapon a decade earlier.

Navy needs differed from the other armed services. Naval warships are big, so they can handle large equipment like free-electron lasers. The navy wanted to be able to adjust the laser's wavelength so the beam could better travel through moist air, which absorbed strongly in the infrared, and free-electron lasers are the only weapon-grade lasers with tunable wavelengths. The navy was also looking to a future of all-electric ships, ruling out rocket-engine lasers, but a perfect fit for the all-electric free-electron laser. In 2007, the Defense Science Board wrote, "Free-electron lasers, with the promise of high power, high beam quality, and wavelength agility, could offer unique advantages for ship-based applications."[29] The panel recommended funding development of free-electron lasers for ship defense as a high priority along with high-powered solid-state and fiber lasers, and improved beam control.[30]

It looked like a solid plan in 2007. The report cited long-term navy plans for free-electron lasers, including starting construction of a one-hundred-kilowatt model in 2010 and demonstrating free-electron laser defense against cruise missiles in 2015. In what was then a far-term plan for 2025, the panel suggested developing and demonstrating a megawatt-class free-electron laser, integrating it with beam control systems, and testing the giant laser at sea.[31]

In early 2009, the navy had plans to issue three contracts intended to lead up to building the one-hundred-kilowatt free-electron laser, with more work to come.[32] But in April 2012, the navy put those plans on the shelf to shift to solid-state lasers, taking advantage of a research program launched a dozen years earlier.[33]

CHAPTER 8

LASER WEAPONS GO SOLID-STATE

Malcolm Wallop, the Senate's biggest laser fan, retired in 1995, but his enthusiasm for laser technology may have rubbed off on someone else in the Capitol. Embedded deep inside the fiscal 2000 defense budget were instructions for the Pentagon to develop a laser master plan that identified potential uses of laser weapons, critical laser technologies, and plans to develop those technologies, as well as providing money to perform the review. The timing was remarkably fortunate.

The end of the Cold War had changed the rules for laser weapons. The space-based laser and the long-planned Zenith Star test in space were still nominally on the books, but going nowhere fast. The Airborne Laser was behind schedule, and questions were being raised about its value. THEL was about to yield some impressive test results on short-range targets, but it, too, had its problems. Big rocket-engine lasers could produce impressive raw power, but they were massive and intricately complex.

Congress wanted a fresh look. The chemical oxygen iodine laser was an advance over the chemical rocket-engine laser but not a dramatic one. It still required pumping gases, mixing chemicals, and complex plumbing. Other laser technologies were advancing much faster. Remarkable progress had been made increasing the power and performance of solid-state lasers using semiconductors, glass, and crystals. The Lawrence Livermore National Laboratory had designed and started building the world's biggest laser at the National Ignition Facility. Civilian laser technology was growing by leaps and bounds. Lasers and fiber optics were fueling the explosive growth of internet capacity, and the telecommunications industry was investing billions of dollars in optics. Lasers were producing ever faster and better electronic chips for computers. Laser surgery had become the hot new way to correct vision. Laser machine tools were cutting, welding, and drilling a wider range of materials. With rocket-engine lasers fizzling out and the chemical oxygen iodine laser proving to be a complex and troublesome beast, it was time to take a fresh look at how the new laser technology being developed in the civilian world might be used in weapons.

THE LASER MASTER PLAN

Congress had made an important statement when it put at the top of its list of requirements for the Department of Defense's new laser plan to consider "potential weapons applications of chemical, solid-state and other lasers."[1] The disappointments of trying to scale glass and ruby lasers and the successful demonstration of rocket-engine gas lasers had pushed solid-state lasers out of the weapon picture in the 1960s. Rocket-engine and other flowing-gas lasers had held the lead in the power race for decades. They could convert over 10 percent of their chemical energy into laser light, and the rapid flow of gas could remove the remaining waste heat, something called the "garbage disposal" principle.[2] Their only rival in the quest for megawatt power at the turn of the millennium was the free-electron laser, which the navy was investigating for its potential use on large ships.

But chemical lasers carried some serious baggage. They were large and vastly complex devices containing hazardous chemicals. If an insurgent rocket were to hit a THEL-like weapon system, it wouldn't just knock out a defensive gun. It would also contaminate the area with hazardous chemicals. The Airborne Laser had to be squeezed to fit into a 747 jumbo jet. And because the Airborne Laser could only carry a limited amount of chemical fuel reserves, it would have to fly home for fresh fuel after tens of shots.[3]

Free-electron lasers got their potential for high power from a beam of relativistic electrons produced by a high-voltage electron accelerator, essentially a specialized type of atom smasher. They could generate kilowatt-class output with good beam quality and turn tens of percent of the electron power into light energy.[4] Free-electron lasers could be tuned in wavelength, which was a big advantage to the navy because it would let them adjust their laser to perform best in the difficult conditions they could encounter at sea.[5] But they were far too massive to get off the ground.

Solid-state lasers had come a long way. A crucial advance came from replacing the lamps that had been the standard source of energy since Theodore Maiman's time with arrays of semiconductor lasers. Semiconductor lasers converted much more of the input electricity into light than ordinary lamps. Their wavelengths also could be matched to the peak absorptions of glass or crystalline lasers, further boosting efficiency. Two kilowatt-class solid-state lasers using this design had been built into the Airborne Laser for illuminating targets and improving beam quality.[6]

The civilian boom fueling rapid improvements in solid-state laser technology had a downside. The high-energy laser report warned that the commercial boom was drawing away the industrial base needed for laser weapons, as Ken Billman had learned the hard way on the Airborne Laser program. A table listed only eleven current suppliers for seven key elements of high-energy lasers, with only five of them considered a "stable business base" for high-energy lasers. Six others were "marginal," among them one of the two makers of optical coatings still producing them for high-energy military systems, Optical Coating Laboratory Inc. (OCLI),[7] which had been bought in early 2000 by a company that wanted to use its coatings for optical communication networks. A few years earlier, the Pentagon had had twenty-four suppliers for those products. It was no wonder they were worried.[8]

The high-energy laser review panel also warned that building big demonstration laser weapons was draining the budget for laser research and development. Some 70 percent of the $227 million spent on high-energy laser science and technology in fiscal 2000 went to big demonstrations like the Airborne Laser and THEL. That left only 30 percent for all other research and development on new technology. Solid-state lasers were tied for first place with the crucial matter of beam control, but each received only $26 million in that year.[9] That would be below the noise level in the hot communications market, where OCLI had sold for a staggering $2.7 billion.[10]

The Pentagon needed to get its high-energy laser act together, the panel concluded. At the top of the panel's list of recommendations was creation of a Joint Technology Office to manage high-energy laser research and development. It should handle day-to-day management and coordinate all the Pentagon's high-energy laser science, including that in the armed services, DARPA, and the Ballistic Missile Defense Organization.[11]

"Programs such as the Airborne Laser (ABL), Space-Based Laser (SBL), and Tactical High Energy Laser (THEL) are desirable to demonstrate that [high-energy laser] weapons can be fielded," the panel wrote, but more money was vital for research.[12] Without more money, the Pentagon couldn't build on what it learned from the big demonstrations, so if the money wasn't available elsewhere, it should be cut from the big demonstrations and shifted to research. The panel strongly urged shifting money from elsewhere in the fiscal 2000 and 2001 budgets to meet research needs.[13]

THE HIGH ENERGY LASER JOINT TECHNOLOGY OFFICE

Initially the High Energy Laser Joint Technology Office (HEL-JTO) included the army, air force, and DARPA; the navy joined later. Its formation in 2000 was the right move at the right time. System engineers were getting a bad feeling about the prospects for high-energy chemical laser weapons.

Engineering is a multifaceted discipline. Electrical engineers specialize in electrical power systems. Electronic engineers specialize in electronics, things like transistors, stereos, phones, computers, and integrated circuits. Mechanical engineers design and build machines; civil engineers design bridges and highways. Chemical lasers are built by chemical engineers who manage the chemistry and aeronautical engineers who manage the gas flow. Often those boundaries overlap, so chemical and aeronautical engineers worked together on rocket-engine lasers.

System engineers integrate the work of other disciplines to make the whole system work. System engineers are the specialists responsible from converting a laboratory experiment into a practical weapon system that soldiers can use on the battlefield. When they started doing that with lasers around the turn of the millennium, they didn't like what they saw.

The Airborne Laser was "a good technology demonstration, and it moved the ball forward in terms of atmospheric compensation, tracking targets, developing models, and doing lethality tests to look at the practicality of laser weapons," says Mark Neice, who was chief of the laser division at the Air Force Research Laboratory until 2004 when he became head of the HEL-JTO. But when the system engineers looked at the support infrastructure at Edwards Air Force Base and White Sands, and inside the cramped quarters of the 747, they did not see a deployable system. The technology base was great, he said, but the systems engineering piece was missing. The story was the same for THEL and would be the same for the Advanced Tactical Laser.[14]

Laboratory demonstrations are cobbled together to show that an idea is feasible. Once an idea passes that test, the system engineers come in to design a system that can be used in the real world. Military systems must be maintained in the field, so technicians must be able to reach the key points. Everything in the Airborne Laser was tightly fitted together. "It was ten pounds of equipment in a five-pound sack. We had stuff squeezed into just about every corner to make it fit," Neice said. With all the chemicals flowing through the plumbing, the engineers took care to install an airtight bulkhead that sealed the flight crew and laser technicians in the front of the 747 safely away from the chemistry flowing through the tubes in the back.

Looking at what needed to be done brought the engineers to a moment of truth. "Everyone collectively said, 'We've got electric boats, electric vehicles, and electric airplanes. Why don't we have an electric laser to run off the available electricity generated in these vehicles?'" Neice recalls.[15]

The answer was obvious. Nobody had ever built a weapon-class solid-state laser. Industrial laser companies had built kilowatt solid-state lasers able to cut through steel sheets close to the laser. Solid-state lasers could produce very intense beams, but they lacked the power to go very far through the air.

HEL-JTO had been created to develop new high-energy laser technology, so they put cranking up the power on solid-state lasers at the top of their project list. They set a goal of one hundred kilowatts to be approached in two steps. The first goal was producing a steady twenty-five-kilowatt beam, with the second to reach the full one hundred kilowatts that would put it in the laser weapons class. They called the project the Joint High Power Solid State Laser, which was given the acronym of JHPSSL (pronounced jay-hip-sul).

The first solicitation did not specify a technology, but all the contractors chose to power laser action in solid slabs of laser material with light from semiconductor lasers.[16] The laser material differed in details, but it all contained neodymium, a rare earth element that absorbs infrared light at 808 nanometers from semiconductor lasers and emits a laser beam at 1,064 nanometers. A broad, thin slab can dissipate waste heat fairly easily from its wide sides and was commonly used for pulsed high-powered solid-state lasers at the time.

Livermore already had one project underway for the army, called the heat capacity laser, using transparent slabs of a ceramic that could be made in larger sizes than crystals. It fired a series of short pulses for about ten seconds until the slab got so hot that it had to be switched out of the laser chamber and cooled down. When that happened, another slab that had already been cooled was switched back in quickly and the cycle started again. In early versions it could fire for seconds before needing to cool down.[17] Livermore tested lamp pumps as well as semiconductor lasers; lamp systems had better beam quality, but the semiconductor laser pump delivered more power.[18] The system reached twenty-five kilowatts but didn't make the cut for the one-hundred-kilowatt test.[19]

Other contracts went to Northrop Grumman and Raytheon, which used different arrangements of exciting slabs with steady beams from banks of powerful semiconductor lasers to reach twenty-five kilowatts by the 2005 deadline.[20]

Meanwhile another solid-state laser technology was coming on fast, in which the laser was essentially an updated version of the first glass laser Elias Snitzer had made in 1961, described in chapter 3. Like that laser, the light-emitting atoms were contained in a central core surrounded by pure glass that kept the light trapped in the core until it emerged through the mirror at the end. However, the glass was stretched into thin and flexible optical fibers, not the millimeter-diameter rod that Snitzer had used. What gave the new fiber lasers their much greater intensity was powering them with light from semiconductor lasers aimed along the length of the fiber. That pump light excited the light-emitting atoms to emit laser light that stayed tightly concentrated in the core until it emerged from the end of the fiber. The resulting fiber laser converted the light from the semiconductor source into a tightly focused laser beam more efficiently than other glass or crystalline lasers could.

In 2004, then HEL-JTO director Ed Pogue called fiber laser progress "very exciting." Separate DARPA-sponsored projects at Southampton Photonics in Britain and IPG Photonics in the United States had both reached kilowatt powers with the high beam quality important for laser weapons. But it remained unclear how much power could be extracted through the tiny end of an optical fiber while maintaining beam quality and avoiding optical damage. Combining light from several fibers might overcome that limit, but nobody was sure how to do it.[21]

Months before the deadline for the twenty-five-kilowatt demonstration, Pogue told me that "solid-state lasers have really changed the thinking on tactical lasers almost completely over the last five to six years." He hoped that within a few years they would be ready for use in test beds like the C-130 laser gunship or MTHEL.[22]

Northrop Grumman and Textron got the nod for the one-hundred-kilowatt round. Northrop Grumman's approach was to build seven lines of parallel slabs, each building up to fifteen kilowatts, then combine the beams by arranging them in a tile pattern to cover a wider aperture that could be focused into a beam. Textron, which acquired the old Avco Everett Research Laboratory laser group when it bought Avco, bounced the laser back and forth between the edges of a larger slab so it followed a zigzag path they called "thin-zag."

DARPA'S LONG SHOT HELLADS

Meanwhile, DARPA had launched a novel laser project called HELLADS, for High Energy Liquid Laser Area Defense System. The concept had come from General Atomics, a San Diego defense company founded in 1955 to study controlled nuclear fusion.[23] Its most famous—or notorious—fusion project was Project Orion, funded by DARPA in 1958 to develop a spacecraft propelled by essentially exploding a series of nuclear bombs behind a giant rocket to push it forward by the pressure from the explosions. Its strength was its sheer audacity and the tremendous energy it could have produced, in theory propelling a four-thousand-ton spaceship to Mars by 1965 and Saturn by 1970. Its fatal flaw was the mind-boggling environmental consequences of both the nuclear fallout of the explosions and the intense pulses of electromagnetic energy that would have blown out terrestrial electrical power networks and electronic systems on the ground and in space.[24]

General Atomics still takes pride in pushing the cutting edge, but its current projects are tamer, including rail guns, the Predator drone, and novel nuclear reactors.[25] HELLADS is its big push on the laser frontier. The idea originated around 2000 as an innovative way to keep high-energy solid-state lasers from melting by extracting the waste heat. "We had a couple of completely new ideas, including a liquid laser. They were considered completely crazy at the time, but DARPA funded us," said project head Mike Perry.[26]

The crucial innovation was carefully matching the optical properties of the solid laser material and of a liquid coolant flowing in channels through the solid. That removed the heat so the laser light could pass undisturbed through both.[27] Flowing liquids through holes in a solid is the fastest way to remove heat from it, but the optical properties of liquids generally don't match those of solids. The key property is the refractive index, which measures how much light bends. If a liquid and a solid have exactly the same refractive index, the light won't bend at all when it passes from one into the other as long as the fluid is flowing very smoothly. That's tricky to do, but it can cool the laser without distorting the light flowing through it, allowing the laser to reach higher power. (It would also make the flowing liquid invisible, as if the whole block was solid glass.)

That intrigued DARPA because they wanted to make laser weapons light and compact enough to fit into fighter jets. Their goal was a laser weapon system able to generate 150 kilowatts that weighed only five kilograms per kilowatt, about 750 kilograms. That's about as much as five large household refrigerators, but it's light by laser weapon standards. If HELLADS could meet that target, it would pack a weapon-class laser into a laser weighing less than 20 percent more than the maximum payload of a large SUV.[28] That would be no lightweight, but it would be an order of magnitude smaller than THEL.

The development program would take years; there were lots of tough nuts to crack. But the potential payoff was big. Fighter pilots would love to have a laser light enough for them to maneuver the jet like a top gun.

DARK HORSE INDUSTRIAL LASERS

Another new technology that came on strong in the early 2000s was industrial fiber lasers.

It might seem strange to get a powerful beam from something as thin and seemingly fragile as an optical fiber, but the fiber geometry has some unique advantages. Glass conducts heat slowly, so a thick slab of glass will stay hot a long time. Heat energy dissipates from the glass at the surface, and it takes a long time for the heat to escape a thick glass casserole, as you may notice when you bake. The thinner the glass, the faster the heat dissipates, and glass fibers are rods that have been stretched into long, thin fibers, so the whole volume is very close to the surface. That makes fibers an attractive shape for getting rid of waste heat before it causes any problems.

Fiber lasers also benefit from how they tightly confine both the light that excites the laser and the light generated by the laser. The closer the input pump light is confined within the laser, the more efficiently it excites the laser, and the more efficiently the pump light is converted into a laser beam. Fiber lasers can be made very efficient by a trick invented by Elias Snitzer in the early 1990s, making glass fibers out of three layers. The innermost layer, the core, contains the light-emitting element, usually ytterbium for fiber lasers (rather than erbium used in fiber amplifiers). The middle layer that surrounds it collects light from the pump sources and confines it so the light passes back and forth through the core as it travels along the fiber. The outer layer confines the pump light in the middle layer.

During the fiber-optic boom of the 1990s, many companies developed fiber lasers to amplify the light in long-distance fiber-optic telecommunications systems. IPG Photonics started in that market, but when it got crowded the company turned to the different business of making high-powered fiber lasers and amplifiers for industrial applications. Founded in 1991 by Russian physicist Valentin Gapontsev, IPG moved its headquarters to the United States in 1998, and in 2000 began producing its own diode lasers to integrate manufacturing and optimize design. Their first lasers started at modest powers of a few watts, but they kept improving them to reach higher and higher powers, reaching kilowatt output in 2002 and ten kilowatts in 2004.[29]

They found a hot market in industrial machining. Their fiber lasers were more compact and efficient and offered higher powers than other solid-state lasers, and IPG became very successful. Their higher powers enabled new applications, and eventually military groups interested in laser weapons realized that they could buy kilowatt-class lasers from industrial firms.

IPG's kilowatt lasers were designed for industrial applications that differed in important ways from what was needed for laser weapons. To achieve high power at low cost, they used fibers with large cores that produced beams that were not confined as tightly as needed for shooting the beam at military targets. All a machine shop needed was to couple the laser beam into a beam delivery fiber that a robotic arm moved near the surface of the material being cut or welded. Laser weapons had to do a different kind of materials-working, on unfriendly objects far from the laser.

That didn't stop military labs and contractors from buying industrial fiber lasers and changing the beam delivery systems for use in field tests of high-energy lasers. Once standard products reached the kilowatt level, it was the simplest and easiest way to get the required power. IPG was happy to sell the lasers, although they didn't publicize their military sales much. The irony of a company founded by Russians selling high-energy lasers to the Pentagon was hard to escape, and the heart of IPG's business was industrial sales.[30]

Kilowatt fiber lasers were far from the ultimate weapon. But if put on a Humvee or other military field vehicle, they could point the beam at an IED or unexploded ordnance a few hundred meters away and wait a little while for the laser to heat the explosive enough to detonate it with a satisfying bang.

The technology also improved in other ways over time. IPG started making single-mode fibers that produced much tighter beams, which could be focused over much longer distances. By 2008, the US Naval Research Laboratory was able to combine the light from four one-kilowatt single-mode IPG lasers into a beam that delivered three kilowatts more than a kilometer away.[31]

AN EXPERT REPORT

Meanwhile, the army's Space and Missile Defense Command commissioned a blue-ribbon panel study of the prospects for solid-state laser weapons for defense against rockets, artillery, and missiles, which had been the main targets of THEL and the proposed MTHEL. It was essentially a reality check in which top scientists review a promising idea.[32]

Most details were classified, but a summary openly released in 2008 revealed considerable caution. The panel warned that the one-hundred-kilowatt-class laser that the army had envisioned had "relatively low technological maturity and relatively high risk, and involves challenging engineering integration." They expected the army would need $100 million more than had been planned to reach their goals. They also thought higher energy was needed and endorsed army plans to try to have a four-hundred-kilowatt laser ready to test by 2018 because it would be a much more effective weapon. The panel estimated that would cost about $470 million.[33]

The proposal showed a caution coming from years of disappointment. The committee met four times in the first half of 2007,[34] which was a bad time for laser weapons. The last old Star Wars laser program, the space-based laser, had been quietly canceled. So had MTHEL. The Advanced Tactical Laser was winding down as system engineers turned thumbs down on chemically powered lasers for their long logistics tails—all the fussy things needed to keep them running. The Airborne Laser was years behind schedule and far over budget. Solid-state lasers were the last big hope for laser weapons.

Their push for a four-hundred-kilowatt laser came largely from a fear that a one-hundred-kilowatt laser might not be enough to kill many targets. Yet setting that goal for 2018 has proven to be overoptimistic. Perhaps they were too far from the field to realize how long it takes to bring such powerful lasers to reality.[35]

The panel tried to rate seven variations on solid-state lasers then in the running for weapon use. Northrop Grumman's JHPSSL design came out on top, although far from ready. DARPA's HELLADS ranked a bit lower because of its high risk and low technological readiness.[36]

Fiber lasers were already successful for laser machining, but the panel ranked them lower because they did not think fiber lasers could provide the needed combination of high power and high beam quality. IPG had reached kilowatt power levels in 2002 and ten kilowatts in 2004 by using fibers with large central cores, a design that increased power but sacrificed beam quality.[37] That was fine for laser machining in a factory but not usable in laser weapons firing at targets from a distance.[38]

What weapons needed was the high-quality, tightly focused beam from fiber lasers with small central cores that operated in what is called a single mode. However, at the time single-mode beams had only reached the kilowatt range, and the panel thought their maximum output would be sharply limited. IPG did demonstrate a single-mode fiber laser that reached ten kilowatts in 2009, but that took some elegant feats of engineering.[39] It looked like the outputs of many fibers would have to be combined to make a high-powered beam, and nobody was sure how to do that.[40]

MAKING GUNPOWDER A TWENTIETH-CENTURY TECHNOLOGY

Both Northrop Grumman and Textron were closing in on reaching the one-hundred-kilowatt JHPSSL goal in early 2009. Both used slabs of transparent ceramic that emitted laser light and conducted heat well but needed liquid cooling to keep the laser operating steadily for the required five minutes. HEL-JTO realized cooling was necessary for the demonstration because only about one-fifth of the electrical energy going into the laser emerged from the laser beam. That meant that the lasers drew five hundred kilowatts of power and needed to remove about four hundred kilowatts of heat to produce a one-hundred-kilowatt beam.[41]

Even before the demonstrations were ready, HEL-JTO launched a follow-up program called the Robust Electrical Laser Initiative (RELI). The goal was to make solid-state lasers robust enough to be used as weapons in the field, rather than being only laboratory test beds. John Wachs, chief of the directed-energy division at the Army Space and Missile Defense Command in Huntsville, Alabama, wanted to push the laser output to 30 percent of the input so the laser would need only 330 kilowatts of input power to generate a one-hundred-kilowatt beam. Then the cooling system would need to only remove 230 kilowatts of heat, a huge reduction. He expected fiber lasers would be in the running as well as better slab lasers.[42] Because cooling equipment is bulky and gobbles electric power—as your electric bills remind you every summer—the win is even bigger.

A Northrop Grumman laser housed in the massive shiny metal box shown in figure 8.1 generated a 105-kilowatt beam for more than five minutes in March 2009, the first to reach the JHPSSL finish line. On a press teleconference, Dan Wildt, the company's vice president of directed-energy systems, joked, "We're doing our part to make gunpowder a 20th century technology."[43] The laser output power they achieved was not quite a fifth of the electrical input, a shade short of the 20 percent ideal, but it was an important achievement. Inside the shiny metal box were seven laser modules, each generating fifteen kilowatts, lined up so they tiled their beams beside each other so they together generated 105 kilowatts. But the shiny metal box measured two by two by 2.7 meters and weighed in at seven metric tons, rather on the hefty side for battlefield use.[44]

Figure 8.1. The Northrop JHPSSL laser could deliver a 105-kilowatt beam for more than five solid minutes, turning a little under 20 percent of the input electrical power into laser energy. It was a breakthrough. But it measured two by two by 2.7 meters and weighed seven metric tons. (Photo courtesy of Northrop Grumman Systems Corporation, Aerospace Systems sector.)

Trade magazine *Aviation Week* recognized the achievement by presenting its Laureate Award to Mark Neice of HEL-JTO; Brian Strickland, chief scientist of the Army Space and Missile Defense Command; and Jay Marno and Stuart McNaught of Northrop Grumman. It was a milestone on the road to electric laser weapons that could keep on firing as long as they have power.[45]

A hundred-kilowatt beam can slice two-inch steel for shipbuilding if focused onto a small spot.[46] But if you measure the power in other ways, it's not quite so impressive. A hundred kilowatts is only equivalent to 137 horsepower, so it can't compete with a muscle car, but it can outclass the 117 horsepower of my 2013 Honda Fit.[47] On the other hand, a car engine's mechanical energy is used over a much larger area than a laser focal spot.

Nor would a high-energy laser make much of a death ray for killing humans. Certainly a focused hundred-kilowatt beam could give a person a nasty burn, but it takes time to deposit energy, so a person likely would jump away in pain before the burn got deep enough to do serious damage to anything other than the eyes. Existing lasers don't have the energy density needed to make handguns. Even if you could find some magical material with enough energy density, a laser that emits half of the electrical power entering it, very good by laser standards, would generate as much heat as it does laser beam power, so Buck Rogers would burn his shooting hand unless he wore very heavy-duty insulated gloves. So gunpowder is still a long way from last-century technology.

Textron's laser came in a few months later, but the Northrop Grumman laser worked better, and HEL-JTO chose that laser to ship to the High-Energy Laser System Test Site at White Sands for tests against targets and to assess how well the beam can travel through the atmosphere. However, it had to be operated in a clean room, and it needed an additional room for water chillers. "The system engineers took a look and said, 'Yeah, but what's that going to fit into?'" recalls Mark Neice, then head of HEL-JTO.[48]

Yet, JHPSSL did demonstrate remarkable progress in increasing laser power. It also offered a tool to test optics and beam propagation with the wavelengths near one micrometer generated by solid-state lasers. It won't find a home on the battlefield, but it can be used for target practice.

ELEPHANTS IN THE LIVING ROOM

At the end of June 2009, I had the chance to get up front and personal with laser weapons when the Directed Energy Professional Society held a meeting on high-energy solid-state lasers at a Boston-area hotel. The biggest laser that I saw on display was the ten-kilowatt single-mode fiber laser from IPG Photonics, the company's latest push into high-energy lasers, notable for its high beam quality and remarkable high power from a fiber with a small core. It was the size of a rather large refrigerator.

For understandable reasons, the society was concerned about having a reporter on the premises, but the classified sessions were held elsewhere, and they were fine after I showed them what I had written about laser weapons.

For the engineers and physicists working on laser weapons—and researchers in allied fields—it was a gathering of old friends and colleagues. I enjoyed that, running into old friends like Martin Stickley, Jack Daugherty, and Louis Marquet whom I had talked with over the years. I chatted with others over lunches and a lobster cookout on the hotel's lawn.

Many of the papers were cutting-edge research that fascinated my inner laser geek. Another friend, Margaret Murnane from the University of Colorado, a leading world expert on short-pulse lasers, gave an opening talk on turning short pulses of ordinary light into ultrafast coherent X-rays by nonlinear processes.[49]

Other papers were weapons oriented. Oleg Shkurikhin of IPG Photonics described development of kilowatt-class fiber lasers spanning narrow ranges of wavelengths that could be combined with other fiber lasers to generate higher powers.[50] That looked like a promising way to get around the limit on power from individual fiber lasers. Alex Yusim of IPG reported breaking the ten-kilowatt barrier for the first time for a single-mode laser. The trick was very carefully engineering the way laser light was produced. The company showed a prototype but cautioned "it will be very difficult to get to 15 to 20 kilowatts."

But the most revealing talk about the real issues in laser weapons was by Sean Ross of the Air Force Research Center, who said that system engineering issues were "the elephant in the living room" of solid-state laser weapons. The three largest concerns were damage to the solid-state laser material at high energies, managing power and heat transfer, and developing rigorous ways to measure laser performance.[51]

Optical damage had put solid-state lasers out of the death-ray business in the 1960s when they shattered glass-laser logs that had been made to produce powerful pulses. Optical damage to the transparent and reflective optics used on rocket-engine lasers was a problem. Now optical damage was "quietly alarming upper management in most high-energy laser programs," Ross said. One program had been delayed several months when optics were damaged at power levels where they were supposed to be safe to use. The cost overrun was a large fraction of a year's operations budget. In another program, little specks of dust caused damage to the optics that took three days to replace and readjust.[52]

"We all damage optics," Ross told a large room full of specialists who expected some damage. "But nobody knows the mean time until failure of high-energy lasers operating at thresholds below the stated damage levels." In short, nobody had a firm handle on an important functional issue. Developers were working on new optical materials, but much work remained.[53]

The latest and greatest demonstration lasers generating kilowatts and up relied on laboratory chillers, totally unsuitable to the military environment. "The bigger systems get, the bigger the cooling requirement," Ross said. "We need to start getting experience in other types of cooling systems," and using non-water coolants, particularly for the air force.

FIELD DEMONSTRATIONS

Well before I attended the Directed Energy Professional Society meeting, I had been hearing of a new generation of laser weapons being tested in the field. They weren't bulky rocket-engine-laser behemoths like the Airborne Laser Laboratory bagging a missile on grainy black-and-white movie film. They started with lasers bolted onto a truck or jeep that were taking out easy targets, like unexploded ordnance or IEDs left exposed on a battlefield or dirt road.

The tests used industrial solid-state or fiber lasers emitting hundreds of watts to several kilowatts. They were standard products available from several companies, compact enough to be put on a vehicle. Add a control system to turn on the laser and aim the beam onto a target, and they were ready to go. The key to success was finding the right type of target. The ideal target was something menacing but motionless and laden with explosives that could be heated to a point where it detonated with a satisfying bang. Those were not common military situations, but the bangs showed that lasers might help in clearing the debris of war, and suggested that lasers were on the way to being useful.

After the USS *Cole* attack, the navy was looking for ways to zap small boats used by insurgents. I got to see the prospects when I visited Bill Shiner, whose latest career was as vice president for sales at IPG's impressive headquarters in central Massachusetts. He took me to a conference room, offering to show me something neat. He started a video, and I saw a small boat with a couple of big outboard engines, bobbing slightly in what looked like shallow water. Nobody was on board, and it may have been anchored or tied down. I knew something was going to happen, I knew a laser was going to be involved, and I knew that the laser beam—like those of most solid-state lasers—was not going to be visible. Soon I saw little flickers of flame around the engine. Outboard motors are messy, and they tend to leak a little gasoline. Within seconds the flames grew and the engine soon caught fire.

You can find that scene on YouTube now, but you couldn't then. It wasn't classified, but back then it was one of those things shown to insiders because it was neat. He didn't have to say where it came from. Because it was a boat on the water, I knew it had the navy's fingerprints.

The Office of Naval Research had plans for a bigger and better test, called the Navy Laser Weapon System, or Navy LaWS. It began with a relatively small industrial laser, which bagged several small boats from a destroyer in 2011, and several unpiloted aircraft in 2012. Then it got an upgrade, and in April 2014 the navy announced it would be deployed on the USS *Ponce* for tests at sea in the Persian Gulf.[54]

The navy didn't say much about the laser then, but soon the word came out that it was built around half a dozen industrial fiber lasers from IPG, each built to emit 5.5 kilowatts. They were built rugged to work on a factory floor, but they probably needed to be more rugged to deal with the moist ocean and bouncing around on a ship. They were hooked to a telescope big enough to make a serious amateur astronomer very happy (see photo insert). The telescope's optics obviously were heavy-duty to handle the beam power that added up to about thirty kilowatts and to deal with the marine environment, which can be harsh. But the navy did nothing elaborate to combine the beams, just fed the light into the telescope through optical fibers.

Figure 8.2. Operators at the Navy LaWS console on the USS *Ponce*. It may look like a video game, but they're shooting real lasers at real targets outside. (Photo courtesy of the Office of Naval Research.)

Inside the ship, the navy installed video screens and a controller, which looked like something from a video game and which was operated with professional calm by a seasoned-looking soldier. It bagged targets including a drone and explosives on a small boat, and to show its success the navy posted a video on YouTube that as of this writing has received over 5.5 million views.[55] Rear Admiral Matthew L. Klunder, chief of naval research, said they ran the prototype "through some extremely tough places and it locked on and destroyed the targets we designated with near-instantaneous lethality." And Klunder added something else that clearly interested the navy. They didn't have to fire any costly interceptor to bag those targets. Nor did they need to carry tank loads of costly and dangerous chemicals. All it needed was diesel fuel, which Klunder estimated cost less than a dollar per shot. At that rate, "there is no question about the value LaWS provides," he said. "With affordability a serious concern for our defense budget this will more effectively manage resources."[56]

THE ROBUST ELECTRIC LASER INITIATIVE

The next key step for HEL-JTO was going beyond the raw power of JHPSSL to make lasers more practical for weapon systems. That meant smaller size and weight, ruggedness, and overall system efficiency higher than 30 percent. They called the challenge the Robust Electric Laser Initiative to stress the importance of robustness. They didn't want any more clean rooms; they wanted a durable, compact, and efficient laser that ran on electric power. They set out to support four different approaches, all due in 2014.

One was General Atomics' HELLADS program, originally launched under DARPA support. That was fundamentally a slab laser project with an important twist, liquid flowing through the slab to remove waste heat without disturbing the light passing through the slab. Its big appeal was the quest for 150 kilowatts from a 750-kilogram laser, small enough to fit into a fighter jet. Pilots are central to the air force culture, so the brass wanted to equip top-gun pilots with their own laser death rays.

A second program was an advanced slab developed by Raytheon with an internal structure called a waveguide, like a long, thin ruler embedded in the slab that guided the laser light along its length. Light is trapped within the waveguide structure so heating can't spread the light out by thermal blooming, a problem in some solid-state lasers.[57] In 2012, Raytheon described a thirty-kilowatt version in which the slab would weigh under two hundred kilograms and occupy less than 0.4 cubic meters, roughly a thirty-inch cube. They claimed the design could be scaled to at least one hundred kilowatts.[58]

Northrop Grumman shifted gears from slabs to combining the outputs from many fiber lasers. The trick is to get the peaks and valleys of the light waves in all of the fiber lasers to match up precisely with the peaks and valleys in all of the others. Obtaining that match requires starting with a single source of stimulated emission, getting all its waves aligned perfectly, then dividing the light between parallel fibers carefully matched so the waves remain aligned together. Ideally, this is inherent in the coherence of the stimulated emission process that generates laser light, but the details are complex. It's fairly easy to get radio transmitters to align up with the peaks and valleys matched because radio waves are centimeters long. Light waves are ten thousand times shorter, making it much harder to match their waves as precisely as needed.

Northrop Grumman started with a single laser source adjusted so it emitted very well controlled light at a narrow range of wavelengths, then split that beam into several parts with the light waves all marching along in perfect phase with each other. Then those beams were routed through separate optical fibers doped with light-emitting elements that amplified the light at that same wavelength, and those beams were very carefully merged together to make a single higher-powered beam. The technique is called "coherent beam combination," and in June 2014 the company reported combining three kilowatt-class outputs to generate 2.4 kilowatts, without any of the beam impairments that had caused problems in earlier efforts to combine multiple fiber laser outputs.[59] Work has continued since then.

Lockheed Martin took a rather different approach. Instead of trying to match all the wave peaks and valleys and wavelengths extremely precisely, they combined light from many fiber lasers emitting at closely spaced wavelengths. Light waves that differ that way don't add together in the same way as light of a uniform wavelength, because the separation of the peaks and valleys changes as the waves move. Instead of interfering with each other's intensity, they just pass through each other and combine to increase total intensity. This works for different colors of light in the air just as it does for radio waves transmitted by radio and television stations broadcasting on different wavelengths.

High-speed fiber-optic communications systems use this principle to send many separate signals through the same glass fibers at different wavelengths. The lasers include slightly different slices of the wavelength spectrum so they don't cause interference, and a hundred or even more wavelengths can be added together. It's called spectral beam combination, and it has been used to combine the outputs of many lasers emitting at slightly different wavelengths—and much greater powers than in communications systems—to generate powers well into the tens of kilowatts.

"Spectral beam combination of high-power fiber lasers can achieve very high combined laser power, with near perfect beam quality . . . for scaling fiber laser technology to weapons levels," says Robert Afzal, Senior Fellow for laser and sensor systems at Lockheed Martin. In 2014, the company used spectral beam combination to blend the output from about ninety-six fiber lasers emitting about three hundred watts each into a single thirty-kilowatt beam for laser weapon trials. In 2017, they shipped a scaled-up sixty-kilowatt version to the Army Space and Missile Defense Command in Huntsville, Alabama, for testing in a mobile demonstrator, a large military field truck.[60] Even higher powers are in the works.

CHAPTER 9

THE QUEST FOR THE ULTIMATE WEAPON

In the years of the incredible laser, the potential of the new device seemed unlimited. It could be the ultimate weapon that Roy Johnson and ARPA had sought to overcome the threat of Soviet ballistic missiles. To doctors the laser could conquer cancer. To physicists it could be the key to understanding atomic physics. To communications engineers it promised a way to transmit almost unlimited amounts of information. All those wonders seemed to be on the horizon, like the exploration of space, which would, inside of a decade, land men on the moon.

Sixty years after Gordon Gould walked into ARPA with the idea of the laser, we've learned nothing is that simple, that good, and that fast. Lasers have enabled a revolution in telecommunications. Pulses of laser light traveling through hair-thin fibers of glass carry words, pictures, voices, video, and vast amounts of computer-generated information around the globe. It's an amazing triumph, but it took forty years, and tremendous advances in electronics and computing as well as in lasers and fiber optics.

Developing laser death rays has proven much more difficult. The only part of the body that laser beams can damage easily is the eye, and doing so deliberately is rightly considered inhumane. Over a hundred countries have ratified the Protocol on Blinding Laser Weapons that bans them.[1] However, the electronic vision systems used by smart bombs and guided missiles are considered fair targets, and laser weapons can target them.

The emerging new generation of solid-state laser weapons can go beyond blinding sensors to damaging, disabling, or destroying enemy weapons. So far they have only been used on test ranges, but the day is coming when they will be taking out drones, rockets, artillery shells, or small boats wielded by insurgents or enemy troops in combat.

APPEALS OF LASER WEAPONS

What attracted ARPA to Gould's proposal in 1959 was the laser's potential to become an ultimate weapon that could deliver energy precisely to a target at the speed of light. In military jargon, lasers are directed-energy weapons, a term that specifies their precision and the fact that they focus energy rather than fire bullets or detonate explosives. Today, what Janet Fender, as chief scientist for the Air Combat Command, and her military colleagues find the most important attraction of laser weapons is their persistence.

Why persistence? "Lasers can stay in the fight longer than expendables," she explains.[2] Modern electrically powered laser weapons run on power from diesel generators that hum constantly on the modern battleground. Bullets, rockets, and bombs can run out, but the laser keeps on firing as long as the diesel holds out and logistics specialists make sure field troops have ample supplies. No commander wants to run out of diesel before the enemy runs out of fight.

"We're not going to junk all our two-thousand-pound bombs," says Fender. The weapons needed depend on the mission. But lasers have the sustained firepower to create a defensive shield to protect an aircraft or a ship, allowing it to stay in the fray longer. That's what field commanders want. "Persistence is the first attribute that everyone agrees on."[3]

Precision lethality is high on the list. "You don't want to be dropping two-thousand-pound bombs on everything," she said. Laser precision minimizes collateral damage and civilian casualties, key concerns in modern counterinsurgency. Laser beams also can be retargeted quickly. Zap one quadcopter drone with a small but menacing payload, and the beam can reach out at the speed of light to zap another target. The precision laser beams also give commanders new options. When he headed the Air Force Special Operations Command, General Bradley Heithold praised the benefits of covert laser operations like reaching out to quietly slice key communications lines without putting troops in harm's way.[4]

Underlying all that is what Fender calls "reversing the cost equation." A sophisticated Patriot missile costs a couple of million dollars. A dollar's worth of diesel fed into a generator will provide enough power for a modern electric laser to zap an insurgent rocket or drone. Today we use expensive weapons to defend against attacks by inexpensive rockets, drones, and small boats. "If we had the low cost per shot capability that lasers offer, then we would be fighting inexpensive threats with inexpensive shots," she says. If enemies attack with more expensive threats, "we would really be reversing the cost equation in our favor, not just equalizing it."[5]

The air force research timetable does not include any laser ultimate weapons. For the near term, the air force is focusing on ground-based applications that seem the most feasible and don't pose technical challenges as severe as trying to squeeze a high-energy laser into a fighter jet. Looking to the longer term, the Missile Defense Agency is studying prospects for nonchemical megawatt lasers that could be installed into super drones for the ballistic missile defense mission once seen for the Airborne Laser.

Fender has seen an important shift recently in how interest grows in laser weapon systems. Historically, developers of laser systems have pushed their new technology to potential military users. Now she sees a pull from users who want new laser technology once they see what lasers can do when tested in an operational environment. For example, the army took demonstration laser systems down to Fort Sill in Oklahoma and put a general officer into the driver's seat along with the weapons officer for laser target shots at drones. "When the quadcopter went up in flames, the general officer said, 'We want a hundred of these,'" Fender says.[6]

A consensus is emerging that laser weapons are now ready to begin moving into operations, she says. "Don't get me wrong. The research is not done, but it's done enough for a first generation."[7]

Lasers and other directed-energy weapons do have important limitations. They can hit their target at the speed of light, but the beam has to dwell for a while to deliver a lethal dose of energy. An interceptor moves like a tortoise compared to a laser beam, but if it hits the target, it can kill it instantly. Watch a laser zap a rocket, and it's not the instant blast you expect from watching science fiction ray guns. Instead, the laser spot glows before it blows. Lock a laser weapon onto a rocket, and the rocket continues moving, with the focal spot getting hotter and brighter, and only after a matter of seconds does the explosive payload reach the detonation point and explode. It's like heating a teakettle of water on an electric stove. Turn on the burner and it warms almost instantly, but it takes time to heat the water to boiling. That means the laser must keep its killer beam focused on the moving target until it explodes or loses control and crashes.

Don't be fooled either by videos showing high-powered laser beams slicing sheet metal almost like butter. That beam is focused across a matter of inches so it can be concentrated on a tiny spot. Laser weapons are hundreds of meters or more from their targets, so the beam can't be focused that tightly onto a spot. It works by heating an area inches across. That can burn a hole through the hood of a stopped truck, but it's better at heating a thin-skinned fuel tank until the pressure blows out the softened metal, or heating an explosive to the detonation point. A laser weapon can ignite gasoline fumes from the tank of an outboard motor, but it won't slice through the heavy armor of an enemy tank on the battlefield.

In any case, lasers do have a special mystique. Dr. Evil in the movie *Austin Powers* didn't just want sharks; he wanted "sharks with frickin' laser beams."[8] US troops overseas now warn off people approaching checkpoints with bright green laser dazzlers, essentially superpowered laser pointers that are so bright that people instinctively turn away from their glare.[9] Using the green lasers is said to have reduced deaths at checkpoints.

LASER WEAPON TESTS AROUND THE WORLD

Today, armies around the world are testing the potential of laser weapons. In the United States, the air force, army, marines, and navy are all evaluating laser weapons with powers from 30 to 150 kilowatts. None are in mass production or considered standard equipment yet, but these laser weapons are being tested in real vehicles out in the real world. That's a big step beyond the laboratory environments and clean rooms needed by the JHPSSL high-energy solid-state laser test bed just a decade ago. Fiber lasers are the hottest technology right now, but HELLADS and Raytheon's waveguide system remain in the running, and new types of lasers are still being developed. It's a field in motion.

Other military powers are also working on solid-state laser weapons in that power range. The Israeli firm Rafael has developed a fiber-laser "Iron Beam" system with undisclosed power to complement the Iron Dome rocket interceptors that Israel has already deployed to stop insurgent rockets.[10] The German defense firm Rheinmetall has demonstrated solid-state laser weapons with output to thirty kilowatts.[11] China's PolyTechnologies is offering a laser weapon system called "Low Altitude Guard II" with output to thirty kilowatts that it claims can shoot down drones up to four kilometers (2.5 miles) away.[12] Britain is spending thirty million pounds sterling to build a fifty-kilowatt laser called "Dragonfire."[13]

The Russians have a long history of developing high-energy lasers and long were the only serious competitors to Pentagon programs. In March 2018, President Vladimir Putin mentioned that the country had "combat laser systems," and Russian deputy defense minister Yuri Borisov later said those systems had been delivered to the armed forces in 2017. An accompanying video shows truck-mounted lasers that appear similar to US demonstration systems.[14]

NAVY LASER TESTS

The US Navy was a leader in testing solid-state laser weapons. Videos showing the Navy LaWS target practice on the USS *Ponce* got a lot of attention. Yet a spokesman for the Office of Naval Research said the thirty-kilowatt shipboard laser "does not have the power to be an operational system," but was useful as a proof of principle and to see how well a laser could work in the difficult marine environment. The spokesman added, "The success on the *Ponce* has been very encouraging for us."[15]

But the USS *Ponce* test was just a first step. The ship, commissioned in 1971, was antiquated, and after tests were finished the laser was moved to another ship and the *Ponce* was decommissioned in 2017.[16] The LaWS had occupied a lot of area on the ship, and for further tests the navy wanted a ship with a better electrical supply to test a new generation of laser.[17] The next test also moved to a different organization, the Naval Sea Systems Command, a system engineering group responsible for integrating new technology into real-world navy systems.[18]

To scale up what they learned from the *Ponce* tests, the navy contracted with Lockheed for a new project called HELIOS, for High Energy Laser with Integrated Optical-dazzler with Surveillance. The heart of the system is a fiber laser delivering 60 to 150 kilowatts. The contract also covers a laser surveillance system to monitor the sky and sea around a ship to spot would-be attackers.[19] Another addition is a green laser "dazzler" that generates a beam too powerful to look at, to blind sensors and obstruct the vision of enemy pilots or small-boat operators who might be attempting an attack. Integrating a dazzler with the high-energy laser gives the defenders an option to confuse or knock out guidance systems before zapping them with the laser weapon.[20]

Lockheed is building two copies of the HELIOS system. The navy will install one on an Arleigh Burke–class guided-missile destroyer and integrate it into the ship's electrical power and cooling systems, as well as its high-tech Aegis battle management system. Those ships are the heart of the operational US destroyer fleet, making the test far more realistic than using the outdated USS *Ponce*. Robert Afzal of Lockheed called it a watershed moment, going from experimental systems to supplying laser weapon capability for ships in the active navy fleet.[21] The second will undergo extensive tests at the High Energy Laser System Test Facility at White Sands. Both HELIOS laser systems use spectral beam combination, which Lockheed has already demonstrated at the thirty and sixty kilowatt levels, that is adapted for use in marine environments. It can convert 35 percent of input electricity into laser power.[22]

AIR FORCE LASER WEAPON PLANS

The Air Force Research Laboratory is testing laser weapons in separate ground and air tests. On the ground, the Demonstrator Laser Weapon System (DLWS) is a joint program with DARPA to integrate the HELLADS laser with the army's ground-based beam control system at White Sands and test how effective lasers can be from the ground against targets including rockets, artillery, mortars, and surface-to-air missiles.[23]

For airborne applications, the lab will test prospects for a compact "medium-power" laser weapon that could be mounted in a pod installable on fighter jets to defend against ground-to-air and air-to-air weapons. The Self-Protect High Energy Laser Demonstrator (SHiELD) will first test target tracking, then flight-test the laser weapon starting in 2020.[24]

LASER-ARMED TANKS AND TRUCKS ON LAND

The US Army Space and Missile Defense Command is testing lasers in standard infantry vehicles. A five-kilowatt laser replacing the heavy gun normally mounted atop an eight-wheeled Stryker shot down over one hundred drones, and Boeing is now installing a ten-kilowatt laser. Size of the laser weapon is limited by the size of the tanklike Stryker.

The army group also is testing bigger lasers in the Heavy Expanded Mobility Tactical Truck (HEMTT) the infantry uses to haul military equipment in the field. They have now replaced the ten-kilowatt industrial laser, originally installed several years ago, with a sixty-kilowatt fiber laser built by Lockheed, using its spectral beam combination technique. The new laser delivers six times more power than the earlier laser, and it fits in the same truck, along with the original beam control system. It converts about 40 percent of the input electricity into laser power, contributing to the smaller size.[25]

MARINE LASER ON A JEEP

The US Marine Corps plans to make room for laser weapons on its new fleet of jeep-like Joint Light Tactical Vehicles (JLTV), which are replacing some of its Humvees. Current versions of the new JLTV fire Stinger missiles, but the marines are developing a laser weapon to use in the future.[26]

ROLES FOR LASER WEAPONS

The new round of field trials beginning around 2020 is a big step up to the 60 to 150 kilowatt class. That's far from an ultimate weapon, but that's where the technology is today, so that's what can be tested.

Pentagon planners expect this generation of laser weapons to be most useful against rockets, artillery, mortars, small boats, and drones no more than a few kilometers away. Those targets carry explosives or fuels that can be detonated by the heat from the laser beam, or have skins or fuel tanks that can be weakened by heat or fractures. Pressurized tanks may explode. Gasoline-powered motors in drones or small boats may leak enough fuel vapor to ignite. But lasers are unlikely to damage a heavily armored tank.

Operationally, planners hope they will get more bangs by investing their bucks in electrically powered lasers, which are cheap to operate, than in costly interceptors. They don't want to use a two-million-dollar Patriot missile to take out a Katyusha rocket that costs a couple of thousand dollars unless it's absolutely necessary. It's hard to sustain a war when you're spending millions of dollars to destroy thousands of dollars of enemy equipment. The math does work for lasers, as long as they are reliable enough that a $10 million laser can fire many thousands of shots.

The highly directional nature of laser light offers another benefit: less collateral damage. Insurgents often make a point of firing rockets from within civilian areas, where a strike at the launcher may kill civilians. Laser beams that miss their targets are not going to explode if they land in a civilian area. That's important from a humanitarian viewpoint and also from a military one. Reducing collateral damage can reduce the number of new insurgents recruited because they were angry about the damage.

Testing is vital to verify laser operation and to look for unexpected effects. A crucial question is how reliable lasers are in the field and how easily they can be serviced. System operation specialists make it clear that a laser mounted in an airplane is not a viable weapon if it has to be flown back home for fresh fuel after thirty shots. Lasers used in the field must be engineered to be repairable or to have field-replaceable modules. Nobody will want a clean room in a battle zone.

BIGGER AND BADDER LASERS

Michael Griffin, the former NASA administrator who was named undersecretary of defense for research and engineering in early 2018, wants bigger and badder lasers for the future. "We need to have 100-kilowatt class weapons on Army theater vehicles. We need to have 300-kilowatt class weapons on Air Force tankers. We need to have megawatt-class directed energy weapons in space for space defense. These are things we can do over the next decade if we can maintain our focus," he told the House Armed Services Committee during an April 17, 2018, hearing. One-hundred-kilowatt lasers are already in testing. He said three-hundred-kilowatt lasers are five to six years away, and a megawatt could be attainable within a decade with persistent investment.[27]

The choice of laser technology for the coming generation of three-hundred-kilowatt lasers is a horse race. "We're looking to scale up the solid state lasers to a level that makes sense for boost phase," said Richard Matlock, director of the advanced technology program at the Missile Defense Agency.[28] The two leading solid-state competitors are fiber lasers and the HELLADS technology from General Atomics. Both technologies look promising for scaling to higher powers by adding modules rather than scaling up the whole structure.[29]

Bigger and badder lasers will be needed to address the need for boost-phase ballistic-missile defense that had been targeted by the Airborne Laser. Pentagon brass decided a lumbering 747 loaded with chemicals and complex plumbing wasn't up for the job, but they still think boost-phase defense can be important. Now the Missile Defense Agency is pursuing a new approach, putting powerful sensors and a higher-energy laser in a large, unpiloted aircraft that could fly well above the clouds.[30] (The power requirement is classified and would depend on how close the robotic attack drone could get to the launch site, but megawatt-class power like the Airborne Laser would be a reasonable guess.)

To test the prospects, the Missile Defense Agency plans a Low-Power Laser Demonstration program that would fly a hundred-kilowatt-class solid-state laser in a high-altitude drone. Key tests will assess how adaptive optics can compensate for atmospheric effects and how well the optical system can focus the laser beam over long distances.[31] A pilotless craft could fly higher than a 747, and higher altitudes would reduce atmospheric effects and might increase laser range. *Aviation Week* reported that bidding documents specified that the drone must be able to fly above sixty-three thousand feet for at least thirty-six hours, and carry loads of 5,000–12,500 pounds.[32] Plans call for flight-testing in 2022 or 2023.[33]

If test results look good, the next step would be building and installing a bigger laser. Nobody is saying how powerful, but calling a hundred-kilowatt laser "low power" suggests the goal would be at least several hundred kilowatts and more likely megawatt class.

MEGAWATT-CLASS LASERS

The power limits for current solid-state and fiber lasers are unknown, but another type of electrically powered laser also is in the running for making megawatt-class lasers. Its unique advantage is its potential for extremely high efficiency. Its oddest feature is a history that goes back to Gordon Gould and his quest to make a laser.

One of the many materials that Gould explored while working on the laser was the alkali metal potassium in its vapor phase. Experimentally, potassium vapor could be nasty stuff. As an alkali metal it has only one electron in its outer electron shell, which made it extremely reactive. But having only a single electron in its outer shell also made it easy to understand the spectrum of potassium and how to excite it to make a laser. All he needed was to pass electrical current into a potassium lamp so it emitted light, and then use light to excite the potassium atoms into a state that would make laser light.

In practice, potassium turned out to be a pain, and nobody got very far with it.[34] Cesium, another alkali metal, seemed to have better prospects, but that also turned out to be troublesome. For a while Gould couldn't work on any of his laser ideas, but after other lasers had been made, the Pentagon quite reasonably declassified cesium because it seemed as if it would never be of any use. That allowed Gould to join two young Technical Research Group physicists, Paul Rabinowitz and Steve Jacobs, working on cesium. Rabinowitz and Jacobs succeeded in making the first cesium laser during the wee hours of a Saturday morning in March 1962, somewhat to Gould's amazement.[35] They published the news, but did nothing more with cesium because it and the other alkali metals were rather a nuisance to work with and rather useless as lasers. In the years that followed, they were forgotten, until Bill Krupke started looking for new types of lasers.[36]

Krupke had worked for decades at Livermore before leaving to consult on laser physics in 1999.[37] He had worked on solid-state lasers for a long time and saw how they could be powered very easily and efficiently with light from semiconductor lasers. But he also realized that solids didn't conduct heat very well, so he started looking for new materials that could be powered by light from semiconductor lasers but would dissipate waste heat better.[38] He found alkali metal vapors should work very well because they absorbed light at wavelengths very close to where they emitted light. Thus all but 1 percent of the light energy absorbed by an atom of the most efficient alkali metal atom, potassium, would emerge as laser light. Not every atom that absorbed light would emit that much energy in the laser beam, but the losses might still be only a few percent.[39]

Livermore, a company called Lasertel, and others are developing these alkali metal vapor lasers. With virtually all work stopped on chemical lasers, the alkali metal lasers, called DPALs, for diode-pumped alkali lasers,[40] are important competitors in developing megawatt-class lasers. However, others note that the overall fraction of energy turned into laser light actually is limited by how efficiently the semiconductor laser turns electric energy into light. If the semiconductor laser converted three-quarters of the input electricity into light, and 95 percent of that light energy was converted into laser light, the actual efficiency would be 71 percent.[41]

Nonetheless, it is seriously ironic that the only type of laser the Pentagon would let Gordon Gould work on without a clearance has become a leading prospect for the very highest power laser weapons. Somewhere, Gould must be chuckling.

FUTURE VISIONS

Superpowered lasers may also have uses other than weapons.

One is to clean up debris that is accumulating in low earth orbit. A major concern about the militarization of space is that if the shooting ever starts, the debris could accumulate to dangerous levels. We think of space as big, but the space in low earth orbit is not all that big, and it's filling up with satellites, leftover booster rockets, and debris from antisatellite tests, breakup of old satellites, and accidental collisions of satellites. In 1978, NASA scientist Donald J. Kessler warned that a space war or a cascade of collisions from fragmentation could eventually fill space with so much shrapnel that space travel became unsafe.[42]

Lasers could remove the space debris from low orbits by focusing enough light onto the objects to slow them down, dropping them into orbits low enough for them to encounter increased air resistance and eventually burn up when they deorbit by falling in the atmosphere. Lasers could do this from orbit or from the ground, with each offering its own advantage. Space-based lasers would have to be launched into space, but they would need less power because they could approach their targets closer.[43] A dedicated space-based kilowatt-class laser powered directly by the sun or by solar power could be used to remove larger objects from orbit, directly deorbiting the smaller ones and dissembling larger ones to remove piece by piece.[44] Much larger and more powerful lasers could be based on the ground and slow down debris as it passed by.

LASER PROPELLED ROCKETS

The idea of using lasers to propel rockets goes back to a 1969 suggestion by Robert L. Geisler of the Air Force Rocket Propulsion Laboratory.[45] His idea was to beam the laser energy to heat a working fluid that would produce thrust instead of a rocket engine. A working group called the idea feasible and predicted, "It should be competitive with or superior to nuclear fission rockets."[46] Arthur Kantrowitz of the Avco Everett Research Laboratory also embraced the idea and proposed launching satellites into orbit with ground-based lasers.[47]

On a closer look, NASA found the numbers daunting. In 1974, Omer Spurlock of the NASA Lewis Research Center calculated that vertical launch of a rocket weighing 100,000 pounds would require 8.7 gigawatts of power, more than twice the power generated by all the hydropower dams along the Colorado River at the time, even ignoring atmospheric and power-conversion losses. It must have been a discouraging finding.[48]

Laser power can be transmitted through the air to provide wireless energy for low-energy aircraft, spacecraft, robots, or other remote vehicles and equipment. In 2006, Jordin Kare, a Livermore laser physicist, founded a power-beaming company called LaserMotive in Seattle to beam laser light to solar cells that would convert the light into electricity. The company won NASA's 2009 Space Elevator Games by showing they could power a five-kilogram robotic climber to ascend a kilometer-long cable with a semiconductor laser and customized solar cell power.[49] The company later teamed with Lockheed Martin to power a small drone that flew for forty-eight hours in a wind tunnel. Later renamed PowerLight Technologies, the company now is working to supply fifty to a thousand watts of power to heavier military drones and their payloads.[50]

LASER ASTEROID DEFENSE AND SPACE TRAVEL

Perhaps the ultimate in laser propulsion is deflecting wayward asteroids. The basic idea is similar to removing space debris from orbit, using laser energy focused onto the surface of an object to push it so its orbit misses Earth. The details depend on the warning time—that is, how far in advance the asteroid is recognized as a potential threat. The longer the lead time, the lower the force needed and the more time it can be applied to change the orbit. With a short warning time, the laser might have to be launched quickly to approach the object closely and apply a strong force over a shorter period of time. Swarms of lasers also could be launched to avoid the need for large power sources such as nuclear reactors.[51]

High-energy lasers are getting so powerful that Philip Lubin of the University of California at Santa Barbara and Gary Hughes of California Polytechnic State University have proposed building an array of solar cell space lasers that could evaporate a whole asteroid half a kilometer across in about a year. In an eerily well-timed press release,[52] they announced their plan the day before a seventeen-meter asteroid exploded over Chelyabinsk, Russia, on February 15, 2013. After the explosion, they said their system would have made short work of the Chelyabinsk object.

Their planned system is called DE-STAR, for Directed Energy Solar Targeting of Asteroids and exploRation. They envision tests with a series of increasingly large arrays of semiconductor lasers starting with a one- and ten-meter array, then scaling up to a ten-kilometer array of lasers that they calculate could vaporize a half-kilometer asteroid as far away as Earth is from the sun. They calculated that much further out in the future light from a thousand-kilometer array could accelerate a ten-ton spacecraft close to the speed of light.

Lubin says the scheme neither requires "technological miracles" nor violates laws of physics. It just requires the power available from a laser to continue increasing as fast as it did over the previous half century. They also assumed future solar cells could convert 70 percent of incident sunlight into electricity and future diode lasers could convert 70 percent of that electric power into laser power. That would require a major increase in solar efficiency, but semiconductor lasers are close to that level in the laboratory today.[53]

Another far-out but fascinating idea is the Breakthrough Starshot, which would use light from a gigantic array of ground-based laser transmitters to launch a fleet of tiny space probes, built on single circuit boards, on a twenty-year journey to Alpha Centauri. A conventional rocket would launch the probes into Earth orbit. Then they would open their solar sails, and, as they came into range above the laser, the beam would switch on at full power. Within two minutes, the laser power would have accelerated the tiny probes to one-fifth of the speed of light and they would be a million kilometers on their way to Alpha Centauri. Upon reaching the star system, they would collect information and images and transmit them via laser beam back to Earth, where the light would arrive over four years later.[54]

A MAGICAL APPEAL

After sixty years, the laser has a magical appeal that goes far beyond the death ray. The Breakthrough Starshot and laser asteroid defense are far-out speculation, but they are based on real science. We can't say they are impossible, but we can't do them today. The discovery of new types of radiation in the late nineteenth and early twentieth centuries attracted similar speculation about death rays. Over time, scientists learned more about radiation and electric power transmission, and what they learned ruled out some of that speculation. Einstein's discovery that nothing can travel faster than the speed of light ruled out some of Tesla's ideas on wireless transmission of electrical power.

What we learn closes some horizons but opens others. Einstein's new physics made the speed of light the ultimate speed limit. But his insight into how radiation interacts with matter opened the door to the laser. The possibility grew discovery by discovery until Charles Townes applied it to make the microwave maser, and he, Gould, and Maiman pushed further to make the laser.

What we know about the laser and radiation shows that some types of science fiction death rays are impossible. No form of radiation can kill people instantly by mere touch, like the death ray of the evil Clutching Hand in the *Exploits of Elaine*. But intense nuclear radiation can kill people more slowly. We also have learned that other kinds of science fiction death rays are possible. High-energy laser beams can "kill" rockets, drones, and small boats in a span of seconds by depositing heat that ignites fuels, detonates explosives, or bursts fuel tanks.

Nonetheless, the laser retains something magical after nearly sixty years.

Arati Prabhakar saw this laser enthusiasm among military officers when she directed DARPA from 2012 to 2017 and found it rather ironic after the troubled history of laser weapons. She worked with lasers as an undergraduate, so she is familiar with them. That experience may make her more realistic.

"Most new technologies are viewed with suspicion and hostility by most military users," she said. "Usually they asked why they should want to change the way they operate to use a new technology. As DARPA director, I spent a lot of time trying to persuade people that fundamental shifts in technology could enhance military capability."[55]

She added, "Laser weapons are the rare counterexample. The enthusiasm of users has for decades outstripped the capability of the technology. The laser is such a powerfully exciting notion that their desire exceeds the capability of the technology. In the area of laser weapons, I spent a lot of time trying to reduce their expectations that lasers would make practical weapons. The desire is immense, but the technological capabilities have always been far behind what people want to have."

She calls the Airborne Laser a great example of the disconnect between expectation and reality. "When a military user imagines a laser weapon, they imagine it's compact enough to put lots of them on board their platform, but that enormous aircraft was just one laser. Lasers have completely changed military capabilities in other areas," she says, citing laser range-finding and target designation, laser radars and sensing, and communications. But so far laser weapons have not.

She has seen enormous advances in laser technology with the shift from gas and solid-state lasers to fiber lasers and their higher efficiency. In the future, she says, "I think laser weapons may be a terrific solution for some very specific applications of enormous value. But there will not be a general purpose laser weapon system to solve all problems because the physics imposes fundamental limits on delivering energy to targets through the atmosphere over long distances."

Despite the limits on laser weapons, she says, DARPA's role in the laser story is a classic one. The agency sponsored early research that had far wider applications than in military systems. "DARPA also sponsored work on microprocessors, the internet, and artificial intelligence for defense needs, and . . . the private sector then seized those technologies to change our society and economy."[56]

Now a four-star general, Ellen Pawlikowski has fond memories of her Airborne Laser days but has learned from the experience. "I'm tough on laser people these days," she says. "It's because they have a reputation of overpromising and under-delivering. The Airborne Laser was one in a long series of examples. It eventually worked, but it took so long and it really needed another round of design for operability and manufacturability, and because of the time it took and the money it cost, the Department of Defense chose not to pursue it." She complains that scientists tend to think that once the basic science is done, getting a laser running is "just an engineering problem."[57] Thinking of Edward Teller using those same words, I have to agree with her.

HOW FAR CAN LASERS GO?

Laser weapons are on a roll on the testing grounds, but they still face engineering challenges. General Pawlikowski, who studied thermodynamics, is worried about heat management. "As long as solid-state lasers are limited to twenty percent efficiency, eighty percent of the power you put into the laser winds up as heat," she says. "My Ellen 'gut sense' based on data is that until we can produce a 150-kilowatt laser with the kind of lethality that we expect for a weapon, we will not field a laser weapon system."[58]

So far, HELLADS is the only solid-state laser said to have reached that power level, and it's still in testing and development.

With Mike Griffin as undersecretary of defense for research and engineering, the laser weapon world has a friend in the Pentagon. "Griffin is a strong advocate for laser defense, but he's a realist. He does not drink the magic Kool-Aid that we can scale a one-kilowatt [alkali laser] to one megawatt in five years," says Jim Horkovich, a veteran laser developer who worked for him at Schafer.[59] Griffin has talked about a need for three-hundred-kilowatt and megawatt-class laser weapons in the future.[60]

Lower-level Pentagon officials speaking at a meeting in September 2017 say much work remains before putting high-energy lasers on the battlefront. "Keep[ing] a laser on a target for a certain amount of time is a challenge," said Colonel Joe Capobianco of the US Army Rapid Capabilities Office. "Lasers are out there [in industry], but as far as military applications, there's still work to be done." He expects lasers to meet military requirements eventually but said the technology is not yet mature enough for war fighting.[61] Rear Admiral Jon Hill, deputy director of the Missile Defense Agency, called lasers a "game-changer" for missile defense but warned that reaching the range and the power levels needed would be costly. "If you want a solution tomorrow, you must invest today."[62]

The really big questions are when and if laser weapons can become what the federal bureaucracy calls a "program of record." That's where it becomes a line item in the budget and has to justify its existence to Congress. That's also where serious money gets committed, and top military careers go on the line. Think of it as a big hurdle, the point where a project is ready to go into production and be put in the hands of war fighters.[63] For a technology to get to that point, it must be ready for use to solve real military problems.

ULTIMATE WEAPONS

Becoming a program of record would not make a weapon system an ultimate weapon. Literally, the ultimate weapon would be a weapon that could overcome all other weapons.

History tells us that no such thing can exist. When the Wright brothers invented the airplane, the horror of the bitter American Civil War was still in living memory. Their invention seemed so miraculous that many people hoped that it would bring an end to war. "With the perfect development of the airplane, wars will be only an incident of past ages," said Edward Burkhart, mayor of Dayton, Ohio, when he presented Wilbur and Orville Wright with medals at a grand celebration of their mastery of flight in 1909.

It was more a wish than a prediction. The brothers sold their first airplane to the US Army in the same year.[64] In June 1917 after aerial bombing had begun and the United States had been drawn into World War I, a saddened Orville Wright wrote,

> When my brother and I built and flew the first man-carrying flying machine, we thought we were introducing into the world an invention which would make further wars practically impossible. That we were not alone in this thought is evidenced by the fact that the French Peace Society presented us with medals on account of our invention. We thought governments would realize the impossibility of winning by surprise attacks, and that no country would enter into war with another of equal size when it knew that it would have to win by simply wearing out the enemy.[65]

By then, Wright had been commissioned a major in the Aviation Section of the Signal Officers Reserve Corps and was working on military aviation. Worried that the war was becoming stalemated with the two sides nearly equal in aviation, he urged the Allies to launch a crash program to build a huge aircraft fleet that could gain control of the air and bomb German munitions plants.[66]

After the war was over, Wright looked back and wrote, "The aeroplane has made war so terrible that I do not believe any country will again care to start a war."[67] Yet soon countries that had weathered the war began seeking "death rays," beams of directed energy that could bring down enemy aircraft, as you read in chapter 1. Engineers and inventors sought to use radio waves, high-voltage discharges, or other devices to stop engines or to stun or kill pilots. None of their devices proved useful, although they indirectly gave rise to radar, the invention that together with the atomic bomb helped bring an end to World War II.

Even after Pearl Harbor, Wright still hoped that airplanes could help restore peace. "It was air power that made such a terrible war possible, but it also is air power that we will have to depend upon to stop it," he wrote in 1943. Later, President Harry Truman presented Wright with the Award of Merit for his service on a national commission on aeronautics. Yet the devastation of the atomic bomb left Wright a deeply saddened man. He wrote to a friend in 1946, "I once thought the aeroplane would end wars. I now wonder whether the aeroplane and the atomic bomb can do it. It seems that ambitious rulers will sacrifice the lives and property of all their people to gain a little personal fame."[68]

Malcolm Wallop was more realistic when he introduced his plan for a fleet of eighteen orbiting chemical-laser battle stations that he said would cover the earth, each with fuel for about a thousand shots and a range approaching three thousand miles (five thousand kilometers). "There is no ultimate weapon, but this one holds great promise for a generation or two of real protection for Americans," he wrote.[69]

Back at the birth of ARPA in 1958, Roy Johnson suggested that his new agency might come up with a death ray, or some kind of ultimate weapon. But enter the words "ultimate weapon" into a search engine, and it will reveal the sad truth. Ultimate weapons are not real. They are the fantastic stuff of fiction, used by superheroes, supervillains, comics, gamers, and the like.

There is no such thing as an ultimate weapon because there is always new and more potent technology to be developed. That's what an arms race is all about, trying to develop a new weapon that is a step beyond the opposition's most potent destructive capability. No weapon can stop the arms race because the other side can always build another weapon as long as it has the means.

Only people can stop an arms race. Ronald Reagan and Mikhail Gorbachev both wanted to ban nuclear weapons when they met in Reykjavik in 1986. What halted them was the specter of Star Wars. Reagan wouldn't give in to Gorbachev's demands to limit SDI to the laboratory. Giving in could have changed the world, but only if their governments would have agreed. In reality, neither government would have agreed to a total denuclearization. Each would have insisted on keeping something of its own as an edge. Congress and the politburo would never have gone along.

Yet within a few years they and their successors did achieve something less unthinkable that had seemed impossible before. They shrank their nuclear arsenals, turning the swords of weapon-grade uranium into the plowshares of reactor fuel. That probably was the best they could have done then. We can only hope to do better in the future.

ACKNOWLEDGMENTS

This book grew from the four decades that I have spent writing about lasers and laser weapons. It's been a fascinating adventure, and over those years many people have spent countless hours talking with me about lasers and laser weapons in person and on the phone. I can't list them all, but I can thank them all for their help, their time, and their friendship.

This book benefits from the work of other writers, and if you want to dig more deeply I can recommend some books I found particularly informative. For the early developments before the laser, I recommend William J. Fanning Jr.'s *Death Rays and the Popular Media: 1876–1939*, Jonathan Foster's *The Death Ray: The Secret Life of Harry Grindell Matthews*, and W. Bernard Carlson's *Tesla: Inventor of the Electrical Age*. I described the birth of the laser in *Beam: The Race to Make the Laser*, but to dig deeper you should read three narratives from different viewpoints—Nick Taylor's *Laser: The Inventor, the Nobel Laureate and the Thirty-Year Patent War*, Charles Townes's memoir *How the Laser Happened*, and Theodore H. Maiman's memoir *The Laser Inventor*. To understand ARPA's origins and early years, I recommend Sharon Weinberger's *The Imagineers of War* and Richard Barber's official history, *The Advanced Research Projects Agency 1958–1974*.

Robert Duffner's *Airborne Laser: Bullets of Light* gives an inside view of the Airborne Laser Laboratory, the most ambitious program of the early rocket-engine laser years from 1970 to the early 1980s, and shows the determination of military scientists and engineers to learn from tough challenges. Donald R. Baucom's *The Origins of SDI: 1944–1983* and my 1984 book *Beam Weapons* (republished *Beam Weapons: The Roots of Reagan's Star Wars*) describe how the idea of space-based laser weapons originated.

Several books cast valuable light on the Reagan administration's controversial "Star Wars" program. Ben Bova's *Assured Survival* makes the case for strategic defense and tells some interesting stories. William J. Broad's *Star Warriors* profiles the young scientists behind the nuclear X-ray laser, and his *Teller's War* describes its downfall. Looking back from 2000, Frances FitzGerald's *Way Out There in the Blue* tells how Star Wars contributed to the end of the Cold War.

Nigel Hey's *The Star Wars Enigma* takes a different view. For an inside view of the Strategic Defense Initiative I recommend *Death Rays and Delusions* by Gerald Yonas, SDI's first chief scientist, who has a keen eye for light moments as well as for serious implications.

The hundred-year archives of *Aviation Week and Space Technology* are the most extensive journalistic chronicles of laser weapon development I know; they are available online to subscribers. The *New York Times* archives lack as much technical detail but have a broader overview and some fascinating accounts of the "death-ray" development in the early twentieth century.

My research also drew heavily on interviews and notes from conferences that I attended over the years. Some interviews came from the excellent oral history archives of the Center for the History of Physics at the American Institute of Physics. Others came from my own archives of interviews conducted for news stories and earlier historical articles. Gordon Gould, William Culver, and Ronald Martin spoke with me in the past but had died before I began this book. I discussed the book with Jerry Pournelle, but he died suddenly before I could arrange an interview. I drew on past interviews with Robert Afzal, George Chapline, Stephen Jacobs, Ed Pogue, John Wachs, and Dan Wildt. Those I interviewed for this book included Ken Billman, Jack Daugherty, J. J. Ewing, Janet Fender, Ed Gerry, Doris Hamill, James Horkovich, Bill Krupke, Louis Marquet, Mark Neice, General Ellen Pawlikowski, Arati Prabhakar, Mike Perry, Howard Schlossberg, Bill Shiner, M. J. Soileau, and Hal Walker. My conference notes come from meetings conducted from the 1980s through 2017 when I attended the Conference on Lasers and Electro-Optics in San Jose.

Linda Greenspan Regan, then at Plenum Press, stimulated my interest in laser weapons when she asked me to write *Beam Weapons* in 1981 and worked with me on it through its publication. I also owe special thanks to Howard Rausch, who hired me to write for *Laser Focus*, and to other magazine editors whom I worked with over the years: Breck Hitz, Jim Cavuoto, Jeff Bairstow, Heather Messenger, Steve Anderson, Conard Holton, Paul Marks, Christina Folz, Stewart Wills, and Amy Nordrum.

I owe special thanks to my agent, Laura Wood of Fine Print Literary Management, for her advice on developing the book and her success in finding a home for it, and to my editor, Steven L. Mitchell at Prometheus Books, for seeing the book's potential and helping me to realize it. My friend and fellow photonics writer, Pat Daukantas, unearthed photos from long ago hidden in strange places to show what laser weapons really looked like. Thanks to Hanna Etu of Prometheus for helping with the publishing details that authors tend to overlook, and to Jeffrey Curry for careful, thoughtful, and helpful copyediting.

Last, but far from least, I am very pleased to thank the Alfred P. Sloan Foundation for a generous grant supporting my research and writing of this book.

NOTES

CHAPTER 1: DEATH RAYS: FROM THUNDER GODS TO MAD SCIENTISTS

1. Wikipedia, s.v. "List of Thunder Gods," last edited August 15, 2018, https://en.wikipedia.org/wiki/List_of_thunder_gods (accessed August 20, 2018).

2. Theodore Maiman, *The Laser Odyssey* (Fairfield, CA: Laser Press, 2000), p. 118.

3. Ernest Volkman, *Science Goes to War* (New York; Chichester: Wiley, 2002), p. 33.

4. Reviel Netz and William Noel, *The Archimedes Codex: How a Medieval Prayer Book Is Revealing the True Genius of Antiquity's Greatest Scientist* (Cambridge, MA: DaCapo, 2007), p. 34.

5. Olivier Darrigol, *A History of Optics from Greek Antiquity to the Nineteenth Century* (Oxford, UK: Oxford University Press, 2012), p. 12.

6. Netz and Noel, *Archimedes Codex*, p. 62.

7. Ibid., p. 34.

8. Ibid., p. 33.

9. D. L. Simms, *Technology and Culture* 18, no. 1 (January 1977): 1–24, http://www.jstor.org/stable/3103202 (accessed February 5, 2018).

10. Jo Marchant, "Archimedes and the 2000-Year-Old Computer," *New Scientist*, December 10, 2008, https://www.newscientist.com/article/mg20026861-600-archimedes-and-the-2000-year-old-computer/ (accessed February 5, 2018).

11. Ibid.

12. G. L. Leclerc de Buffon, *Memoires de l'Academie Royale des Sciences pour 1747* (Paris, 1752), pp. 82–101; cited in Klaus D. Mielenz, "Eureka!" *Applied Optics* 13, no. 2 (February 1974): Al4 & Al6; the incident is also mentioned in Trevor I. Williams, ed., *A Biographical Dictionary of Scientists*, 3rd ed. (New York: Halsted, 1982), p. 88.

13. *Washington Star-News*, November 13, 1973, p. A3 and *Time*, November 26, 1973, p. 60; cited in Mielenz, "Eureka!"

14. *MythBusters*, episode 16, "Ancient Death Ray," directed and written by Peter Rees, originally aired on September 29, 2004, https://mythresults.com/episode16 (accessed May 4, 2018).

15. "Archimedes Death Ray," 2.009 Product Engineering Processes, October 2005, http://web.mit.edu/2.009/www/experiments/deathray/10_ArchimedesResult.html (accessed May 4, 2018).

16. "2.009 Archimedes Death Ray: Testing with MythBusters," 2.009 Product Engineering Processes, October 21, 2005, http://web.mit.edu/2.009/www/experiments/deathray/10_Mythbusters.html (accessed May 4, 2018).

17. *MythBusters*, episode 46, "Archimedes' Death Ray Revisited," produced by Richard Dowlearn, originally aired on January 25, 2006, https://mythresults.com/episode46 (accessed May 4, 2018).

18. H. Bruce Franklin, *War Stars: The Superweapon and the American Imagination* (Oxford: Oxford University Press, 1988), p. 21.

19. William J. Fanning Jr., *Death Rays and the Popular Media, 1876–1939: A Study of Directed Energy Weapons in Fact, Fiction, and Film* (Jefferson, NC: McFarland, 2015), pp. 27–28.

20. Ibid.

21. Neil Baldwin, *Edison: Inventing the Century* (New York: Hyperion, 1995), pp. 200–202.

22. W. Bernard Carlson, *Tesla: Inventor of the Electrical Age* (Princeton, NJ: Princeton University Press, 2013), pp. 17–18.

23. Ibid., pp. 34–73.

24. Ibid., pp. 117–25.

25. Fanning, *Death Rays*, p. 30.

26. Ibid.

27. Richard Moran, *Executioner's Current: Thomas Edison, George Westinghouse, and the Invention of the Electric Chair* (New York: Knopf Doubleday, 2007), pp. xxi–xxii.

28. Carlson, *Tesla*, p. 222.

29. "Nikola Tesla Discusses X Rays," *New York Times*, March 11, 1896, p. 16.

30. Ibid.

31. Henry C. King, *The History of the Telescope* (Mineola, NY: Dover, 1979), pp. 140–41.

32. Steven Beeson and James W. Mayer, *Patterns of Light: Chasing the Spectrum from Aristotle to LEDs* (New York; London: Springer, 2008), p. 149.

33. Walter Gratzer, *The Undergrowth of Science* (Oxford: Oxford University Press, 2000), pp. 1–28.

34. Fanning, *Death Rays*, p. 31–32.

35. Ibid.

36. Ibid.

37. Philip F. Schewe, "Laser Lightning Rod," *Physics Today* 58, no. 2 (February 2005): 9, https://physicstoday.scitation.org/doi/10.1063/1.4796869 (accessed May 7, 2018).

38. Matteo Clerici et al., "Laser Assisted Guiding of Electrical Discharges around Objects," *Science Advances* 1, no. 5 (June 19, 2015), http://advances.sciencemag.org/content/1/5/e1400111.full (accessed May 6, 2018).

39. Carlson, *Tesla*; Baldwin, *Edison*.

40. Carlson, *Tesla*, pp. 216–17.

41. Ibid., pp. 225–31.

42. Ibid.

43. Ibid., pp. 259–61.

44. Ibid.

45. Ibid., p. 250, citing "Tesla's System of Electric Power Transmission through Natural Media," *Electrical Review* 13 (October 26, 1898): 124–26.

46. Carlson, *Tesla*, pp. 256–59.

47. Ibid., p. 265.

48. Ibid., p. 300.

49. Ibid., pp. 311–30.

50. Ibid., pp. 346–63.

51. Ibid., pp. 364–67.

52. Fanning, *Death Rays*, pp. 134–36; refers to George Griffith, *The World Masters* (London: John Long, 1903).

53. "From Day to Day," *West Australian Sunday Times*, August 7, 1898, p. 2, cited in Fanning, *Death Rays*, p. 34.

54. H. G. Wells, *The Invisible Man and The War of the Worlds* (New York: Washington Square Press, 1965), p. 174 (start of chap. 6).

55. Fanning, *Death Rays*, pp. 40–48.

56. "Invention of an Italian May Put an End to War; Guilio Ulivi Has Detonated Explosives at a Distance of Several Miles by Using Infra-Red Rays and Says World's Fleets Are at the Mercy of His Apparatus," *New York Times*, June 21, 1914, p. 48.

57. "Inventor Elopes on Eve of Tests," *New York Times*, July 18, 1914, p. 1.

58. "Calls Ulivi Bomb a Chemical Fake," *New York Times*, July 20, 1914, p. 1.

59. Fanning, *Death Rays*, pp. 47–48.

60. Wikipedia, s.v. "Craig Kennedy," last edited June 14, 2018, https://en.wikipedia.org/wiki/Craig_Kennedy (accessed August 21, 2018).

61. Arthur B. Reeve, "Chapter 9: The Death Ray," in "*The Exploits of Elaine*," Classic Reader, http://www.classicreader.com/book/1781/9/ (accessed February 5, 2018); *The Exploits of Elaine*, directed by Louis J. Gasnier, Whartons Studio, 1914.

62. Wikipedia, s.v. "The Exploits of Elaine," last edited May 6, 2018, https://en.wikipedia.org/wiki/The_Exploits_of_Elaine (accessed August 21, 2018).

63. Neil G. Caward, "Playing Hide and Seek with Death," *Motography* 13 (March 6, 1915): 351.

64. Robert Buderi, *The Invention That Changed The World* (New York: Simon & Schuster, 1996), p. 52.

65. Eugene Debeney, "The War of Tomorrow," *New York Times*, September 25, 1921, p. 80, https://timesmachine.nytimes.com/timesmachine/1921/09/25/98744919.html?pageNumber=80 (accessed February 9, 2018).

66. Fanning, *Death Rays*, pp. 59–66.

67. Fanning, *Death Rays*, p. 57; quoting "Science Will Win Next War," *Agitator* (Wellsboro, PA), August 4, 1920, p. 6.

68. *Electrical Experimenter*, February 1919, http://www.electricalexperimenter.com/n10electricalexperi06gern.pdf (accessed May 7, 2018).

69. Wikipedia, s.v. "Hugo Gernsback," last edited July 27, 2018, https://en.wikipedia.org/wiki/Hugo_Gernsback (accessed August 21, 2018).

70. *Encyclopedia of Science Fiction*, 3rd ed., s.v. "Science and Invention," March 27, 2017, http://sf-encyclopedia.com/entry/science_and_invention (accessed May 7, 2018).

71. "Milestones: First Operational Use of Wireless Telegraphy, 1899–1902," Engineering and Technology Wiki, last modified December 31, 2015, http://ethw.org/Milestones:First_Operational_Use_Of_Wireless_Telegraphy,_1899-1902 (accessed February 9, 2018).

72. Jonathan Foster, *The Death Ray: The Secret Life of Harry Grindell Matthews* (Inventive Publishing, 2008), pp. 22–24.

73. Foster, *Death Ray*, pp. 26–58.

74. Foster, *Death Ray*, pp. 60–72.

75. Fanning, *Death Rays*, p. 67.

76. Martin Gilbert, *Winston S. Churchill*, vol. 5, *The Prophet of Truth: 1922–1939* (Boston, MA: Houghton Mifflin, 1977), p. 50: quoted in Fanning, *Death Rays*, p. 68.

77. Wikipedia, s.v. "Electronic Warfare," last edited August 20, 2018, https://en.wikipedia.org/wiki/Electronic_warfare (accessed August 22, 2018).

78. Foster, *Death Ray*, p. 226.

79. "Tells Death Power of 'Diabolical Rays,'" *New York Times*, May 21, 1924, pp. 1, 3.

80. Foster, *Death Ray*, p. 127.

81. "Tells Death Power of 'Diabolical Rays."

82. Ibid.

83. Samuel McCoy, "'Diabolic Ray' Makes Scientists Wonder," *New York Times*, June 1, 1924, p. 159, https://timesmachine.nytimes.com/timesmachine/1924/06/01/101600495.html (accessed February 10, 2018).

84. Ibid.

85. "The 'Death Ray' Rivals," *New York Times*, May 29, 1924, p. 18, https://timesmachine.nytimes.com/timesmachine/1924/05/29/104038845.html?pageNumber=18 (accessed February 10, 2018).

86. McCoy, "'Diabolic Ray.'"

87. Avram Balabanovic, "The Electric Wars: Tesla vs. Putin," *Britic*, November 23, 2011, http://www.ebritic.com/?p=139858 (accessed May 8, 2018).

88. "The Death Ray: Harry Grindell Matthews," Harry Grindell-Matthews, Pathe Exchange, 1924, YouTube video, 8:13, https://www.youtube.com/watch?v=qbNgvHfK4wI (accessed February 11, 2018).

89. "War's Latest Terror!" Harry Grindell-Matthews, copyrighted by Pathe Exchange, 1924, YouTube video, 0:55, https://www.youtube.com/watch?v=4IpLjyKSZRw (accessed February 11, 2018).

90. "Hurt by Death ray, Inventor Aids Cure," *New York Times*, July 30, 1924, p. 15, https://timesmachine.nytimes.com/timesmachine/1924/07/30/99453674.html?pageNumber=15 (accessed February 10, 2018).

91. Fanning, *Death Rays*, p. 72, citing "The Versatile Ray, from Mechanics to Medicine, Lord Birkenhead's Comments," *The Times* (of London), May 28, 1924, p. 15.

92. Fanning, *Death Rays*, pp. 56–81.
93. Foster, *Death Ray*, pp. 189–92.
94. Carlson, *Tesla*, p. 367.
95. Ibid., p. 379.
96. Ibid.
97. Ibid., p. 389.
98. "Tesla's New Device Like Bolts of Thor," *New York Times*, December 8, 1915, p. 8, https://timesmachine.nytimes.com/timesmachine/1915/12/08/104659302.html?pageNumber=8 (accessed February 11, 2018).
99. Nikola Tesla, "World System of Wireless Transmission of Energy," *Telephone and Telegraph Age*," October 16, 1927, available online at http://www.tfcbooks.com/tesla/1927-10-16.htm (accessed February 12, 2018).
100. Carlson, *Tesla*, p. 380.
101. Orrin E. Dunlap Jr., "An Inventor's Seasoned Ideas," *New York Times*, April 8, 1934, p. 160, https://timesmachine.nytimes.com/timesmachine/1934/04/08/93759286.html?pageNumber=160 (accessed February 12, 2018).
102. "Tesla, at 78, Bares New 'Death-Beam.' Invention Powerful Enough to Destroy 10,000 Planes 250 Miles Away, He Asserts," *New York Times*, July 11, 1934, p. 18, https://timesmachine.nytimes.com/timesmachine/1934/07/11/93633178.html?pageNumber=18 (accessed February 12, 2018).
103. Ibid.
104. Carlson, *Tesla*, pp. 381–84.
105. Robert Buderi, *The Invention That Changed the World* (New York: Touchstone/Simon & Schuster, 1996), p. 54.
106. Ibid., pp. 53–55.
107. Ibid.
108. Ibid.
109. Ibid.
110. "Electro-Tank Shoots Lightning Rays," *Modern Mechanix*, August 1935, p. 81, http://blog.modernmechanix.com/electro-tank-shoots-lightning-rays/#more (accessed February 12, 2018).
111. K. T. Compton, L. C. Van Atta, and R. J. Van de Graaff, "The Van de Graaff Generator," in *Progress Report on the MIT High-Voltage Generator at Round Hill* (Cambridge, MA: MIT Institute Archives & Special Collections, December 12, 1933), https://libraries.mit.edu/_archives/exhibits/van-de-graaff/ (accessed May 9, 2018).
112. Foster, *Death Ray*, pp. 214–19.
113. Sava Kosanovic, "Ex-Yugoslav Aide, Ambassador to US '46–50, Is Dead-Former Minister of Information and State Delegate to Paris Talks," *New York Times*, November 15, 1956, p. 35, http://www.nytimes.com/1956/11/15/archives/sava-kosanovic-exyugoslav-aide-ambassador-to-us-4650-is-deadformer.html (accessed February 12, 2018).

114. Carlson, *Tesla*, pp. 390–94.

115. John G. Trump to Walter Gorsuch, alien property custodian, January 30, 1943.

116. Ibid.

117. William Thomas, "A Profile of John Trump, Donald's Accomplished Scientist Uncle," *Physics Today*, http://physicstoday.scitation.org/do/10.1063/PT.5.9068/full/ (accessed February 12, 2018).

118. Trump, letter to Gorsuch.

119. Ibid.

120. Walter E. Grunden, "*Secret Weapons & World War II: Japan in the Shadow of Big Science*" (Lawrence, KS: University Press of Kansas, 2005), p. 94.

121. Albert Speer, "*Inside the Third Reich*" (New York: Macmillan, 1970), p. 464.

122. Grunden, "*Secret Weapons & World War II*", p. 114.

123. H. Tsien et al., *Technical Intelligence Supplement: A Report of the AAF Scientific Advisory Group* (Wright Field, Dayton, OH: Headquarters Air Materiel Command, Publications Branch, Intelligence T-2, 1946), p. 159.

124. See, for example, the compilation of covers at the Magazine Art site, http://www.magazineart.org/main.php (accessed May 9, 2018).

125. John M. Miller, "*Murder in the Air*," directed by Lewis Seiler (Turner Classic Films, 1940), http://www.tcm.com/this-month/article/218498%7C0/Murder-in-the-Air.html (accessed May 9, 2018).

126. Harold "Hal" Walker, in interviews with the author, July 18, 2017 and April 18, 2018; "Hildreth 'Hal' Walker Jr.," Historical Inventors, Lemulson-MIT, https://lemelson.mit.edu/resources/hildreth-%E2%80%9Chal%E2%80%9D-walker-jr (accessed May 9, 2018).

127. Walker, interviews with the author; "Hildreth 'Hal' Walker Jr."

128. Ibid.

129. Ibid.

130. "Hildreth 'Hal' Walker Jr."

131. Christopher Klein, "The Birth of Satellite TV, 50 Years Ago," History Channel, July 23, 2012, https://www.history.com/news/the-birth-of-satellite-tv-50-years-ago (accessed May 9, 2018).

132. *The War of the Worlds*, directed by Byron Haskin, Paramount Pictures, 1953.

CHAPTER 2: HOW THE PENTAGON ALMOST INVENTED THE LASER

1. Nick Taylor, *Laser: The Inventor, the Nobel Laureate, and the 30-Year Patent War* (New York: Simon & Schuster, 2000), pp. 13–19.

2. Arnie Heller, "Laser Technology Follows in Lawrence's Footsteps," *Energy and*

Technology Review, Lawrence Livermore National Laboratory, May 2000, https://str.llnl.gov/str/Hargrove.html (accessed May 12, 2018).

3. Taylor, *Laser*, pp. 20–26.
4. Ibid.
5. Ibid., pp. 27–37.
6. Taylor, *Laser*, pp. 52–55.
7. Charles Townes, *How the Laser Happened: Adventures of a Scientist* (Oxford: Oxford University Press, 1999), pp. 49–50.
8. Taylor, *Laser*, p. 57.
9. Ibid., pp. 53–54.
10. Robert Bird, personal communication with the author, July 2003.
11. Townes, *How the Laser Happened*, p. 66.
12. Charles H. Townes, "Masers," in *The Age of Electronics*, ed. Carl J. Overhage (New York: McGraw-Hill, 1962), p. 166.
13. William West, "Optical Pumping," in *McGraw-Hill Encyclopedia of Science and Technology*, 8th ed., vol. 12 (New York: McGraw-Hill, 1997), p. 467.
14. "The Nobel Prize in Physics 1966: Alfred Kastler," Nobel Institute, https://www.nobelprize.org/nobel_prizes/physics/laureates/1966/ (accessed February 15, 2018).
15. Gordon Gould, interview with author for *Omni* (magazine), 1983, unpublished.
16. Taylor, *Laser*, pp. 57–59.
17. Ibid., pp. 60–61.
18. Ibid., pp. 64–65.
19. Townes, *How the Laser Happened*, pp. 64–68.
20. Taylor, *Laser*, p. 18.
21. Ibid., pp. 31, 33.
22. Ibid., pp. 64–67.
23. Gould, interview with author.
24. Taylor, *Laser*, pp. 67–70.
25. Townes, *How the Laser Happened*, p. 96.
26. Ibid., p. 93.
27. Taylor, *Laser*, pp. 71–73.
28. Ibid., pp. 84–85.
29. Ibid., pp. 86–87; Spelling of Holbrook's name from United States Army, *History of Strategic Air and Ballistic Missile Defense*, vol. 2, *1956-1972* (Washington, DC: US Army Center of Military History, 1975), p. 187, https://history.army.mil/html/books/bmd/BMDV2.pdf (accessed May 10, 2018).
30. Taylor, *Laser*, pp. 87–88.
31. US Army, *History of Strategic Air and Ballistic Missile Defense*, p. 186.
32. Annie Jacobsen, *The Pentagon's Brain: An Uncensored History of DARPA, America's Top Secret Military Research Agency* (New York: Little, Brown, 2015), p. 6.

33. James Baar, "Death Ray Visualized as H-Bomb Successor," *Washington Post*, May 6, 1958, p. A-8.

34. Richard J. Barber, *The Advanced Research Projects Agency, 1958–1973* (Washington, DC: Department of Defense, December 1975), p. I-6, and adjacent material.

35. Ibid., pp. II-1–22.

36. A. L. Schawlow and C. H. Townes, "Infrared and Optical Masers," *Physical Review* 112, no. 6 (December 15, 1958).

37. Lawrence Goldmuntz, oral history interview by Joan Bromberg, October 21, 1983, American Institute of Physics, https://www.aip.org/history-programs/niels-bohr-library/oral-histories/4633 (accessed December 29, 2016).

38. Taylor, *Laser*, pp. 86–88.

39. Ibid., p. 88.

40. Ibid.

41. Gould, interviews with author, 1983 and September 19, 1984.

42. Taylor, *Laser*.

43. Ibid.

44. Gould, interview with author, 1983.

45. Taylor, *Laser*.

46. Goldmuntz, interview by Bromberg.

47. Taylor, *Laser*.

48. Townes, *How the Laser Happened*, p. 103.

49. Data from 1959 in Bureau of the Census, "Average Income of Families up Slightly in 1960," *Current Population Reports: Consumer Income*, series P-60, no. 36 (June 9, 1961).

50. Taylor, *Laser*, p. 89.

51. Goldmuntz, interview by Bromberg.

52. Taylor, *Laser*, p. 90.

53. Gould, in interview with author, in Jeff Hecht, *Laser Pioneers*, rev. ed. (Boston: Academic Press, 1991), p. 118.

54. US Army, *History of Strategic Air and Ballistic Missile Defense*.

55. James A. Fusca, "ARPA Seeks 1970–80 Missile Defenses," *Aviation Week*, March 2, 1959, p. 18.

56. Barber, *Advanced Research Projects Agency*, p. III-49.

57. Sharon Weinberger, *The Imagineers of War* (New York: Knopf, 2017), p. 51.

58. Ibid., p. 66.

59. Taylor, *Laser*, p. 92.

60. Townes, *How the Laser Happened*, p. 103.

61. Taylor, *Laser*, pp. 92–95.

62. Taylor, *Laser*, p. 92.

63. Jessica Wang, *American Science in an Age of Anxiety* (Chapel Hill: University of North Carolina Press, 1998).

64. Theodore Maiman, *The Laser Odyssey* (Fairfield, CA: Laser Press, 2000), p. 40.

65. Taylor, *Laser*, pp. 92–96.

66. Ibid., pp. 96–97.

67. Jeff Hecht, *Beam: The Race to Make the Laser* (New York: Oxford University Press, 2005), pp. 78–81.

68. D. F. Nelson, "Reminiscence of Schawlow at the First Conference on Lasers," in *Lasers, Spectroscopy, and New Ideas: A Tribute to Arthur L. Schawlow*, ed. W. M. Yen and M. D. Levenson (Berlin: Springer-Verlag, 1987), pp. 121–22.

69. Gould, in interview with author in Hecht, *Laser Pioneers*, p. 120.

70. Ibid.

71. Ibid.

72. "Anti-Missile Ray Is Pressed by US," *New York Times*, July 5, 1959, p. 8.

73. Hecht, *Beam*, pp. 60–72, 83–94.

74. William R. Bennett Jr., "Background of an Invention: The First Gas Laser," *IEEE Journal on Selected Topics in Quantum Electronics* 6 (Nov/Dec 2000): 869–75.

75. Gordon Gould, oral history interview by Joan Bromberg, session 2, October 23, 1983, American Institute of Physics, https://www.aip.org/history-programs/niels-bohr-library/oral-histories/4641-2 (accessed December 29, 2016).

76. Hecht, *Beam*, pp. 121–23.

77. Stephen Jacobs, telephone interview with the author, June 18, 2001.

78. Gould, interview with author, 1983.

79. Taylor, *Laser*, p. 113.

80. Townes, *How the Laser Happened*, pp. 134–39.

81. Taylor, *Laser*, pp. 121–24.

82. F. A. Butayeva and V. A. Fabrikant, *Research in Experimental and Theoretical Physics*, Memorial Volume in Honor of G. S. Landsberg, trans. Bela Lengyel (Moscow: USSR Academy of Sciences Press, 1959), copy of pp. 1–13 supplied by Colin Webb. Other pages are missing.

83. "The Nobel Prize in Physics 1964," Nobel Institute, https://www.nobelprize.org/nobel_prizes/physics/laureates/1964/ (accessed August 27, 2018).

84. Taylor, *Laser*, pp. 235–58.

85. Arthur L. Schawlow, "From Maser to Laser," in *Impact of Basic Research on Technology*, ed. Behram Kursunoglu and Arnold Perlmutter (New York: Plenum, 1973), pp. 113–48.

86. Maiman, *Laser Odyssey*, pp. 17–35.

87. Hecht, *Beam*, pp. 106–15, 147–49.

88. Ibid., pp. 147–51.

89. Ibid., p. 113.

90. Ibid., pp. 147–57.

91. T. H. Maiman, "Optical and Microwave-Optical Experiments in Ruby," *Physical Review Letters* 4 (June 1, 1960): 564–66.

92. This section is based on Hecht, *Beam*, pp. 3–6 and pp. 169–89; Maiman, *Laser*

Odyssey, pp. 97–107. Sources for the *Beam* account also include interviews with Irnee D'Haenens, Robert Hellwarth, Bela Lengyel, and George Birnbaum, and an interview with George Smith in the archives of the Center for the History of Physics at the American Institute of Physics.

93. Maiman, *Laser Odyssey*, pp. 109–12; T. H. Maiman, "Stimulated Optical Radiation in Ruby," *Nature 187*(1960): 493.

94. Hecht, *Beam*, pp. 3–6 and pp. 169–89.

95. Maiman, *Laser Odyssey*, p. 118.

96. Ibid.

97. Maiman, *Laser Odyssey*, p. 118.

98. Ronald Martin, interview with the author, January 4, 2002.

99. Ibid.

100. Maiman, *Laser Odyssey*.

101. Martin, interview with author.

102. Walter Sullivan, "Air Force Testing New Light Beam," *New York Times*, October 14, 1960, p. 16.

103. Richard Smith, telephone interview with the author, July 5, 1994.

CHAPTER 3: THE INCREDIBLE ROCKET-ENGINE LASER

1. John F. Kennedy, "The Decision to Go to the Moon," speech before a Joint Session of Congress, May 25, 1961, NASA History Office, https://history.nasa.gov/moondec.html (accessed May 14, 2018).

2. Stuart H. Loory, "The Incredible Laser: Death Ray or Hope," *This Week* (magazine), November 11, 1962, pp. 7–8, 25.

3. Jeff Hecht, *Beam Weapons: The Next Arms Race* (New York: Plenum, 1984), p. 26.

4. C. Martin Stickley, in interview by Robert W. Seidel, Niels Bohr Library & Archives, American Institute of Physics, College Park, MD, September 22 1984, www.aip.org/history-programs/niels-bohr-library/oral-histories/4905 (accessed August 30, 2018).

5. C. Martin Stickley, "The Shift of Optics R&D Funding and Performers over the Past 100 Years," *OSA History of Optics*, ed. Paul Kelley, Govind Agrawal, Mike Bass, Jeff Hecht, and Carlos Stroud (Washington, DC: The Optical Society, 2015), p. 186.

6. Stickley, interview by Seidel.

7. Walter Sullivan, "Air Force Testing New Light Beam," *New York Times*, October 14, 1960, p. 16.

8. Robert W. Seidel, "How the Military Responded to the Laser," *Physics Today* 41, no. 10 (October 1988): 36–43; 38; https://doi.org/10.1063/1.881156 (accessed August 30, 2018).

9. Quoted in Robert W. Seidel, "How the Military Responded to the Laser," *Physics Today*, October 1988, p. 36–43; 37n17.

10. William R. Bennett Jr., "Background of an Invention: The First Gas Laser," *IEEE Journal of Selected Topics in Quantum Electronics* 6 (November–December 2000): 869–75.

11. L. F. Johnson and K. Nassau, "Infrared Fluorescence and Stimulated Emission of Nd+3 in CaWO4," *Proc. IRE* 49, no. 11 (November 1961): 1704–705.

12. E. Snitzer, "Optical Maser Action of Nd+3 in Barium Crown Glass," *Physical Review Letters* 7, no. 12 (1961): 444–46.

13. Elias Snitzer, oral history interview by Joan Bromberg, August 6, 1984, American Institute of Physics, https://www.aip.org/history-programs/niels-bohr-library/oral-histories/5057 (accessed May 14, 2018).

14. Ibid.

15. Jeff Hecht, "The Amazing Optical Adventures of Todd-AO," *Optics & Photonics News* 7, no. 10 (1996): 34–40, https://doi.org/10.1364/OPN.7.10.000034 (accessed May 14, 2018).

16. Snitzer, interview with Bromberg.

17. Ibid.

18. Ibid.

19. Ibid.

20. Ibid.

21. Robert W. Seidel, "From Glow to Flow: A History of Military Laser Research and Development," *Historical Studies in the Physical and Biological Sciences* 18, no. 1 (1987): 120.

22. Ibid.

23. William Culver, in telephone interview with the author, October 16, 2008.

24. Ibid.

25. Ibid.

26. Ibid.

27. Seidel, "From Glow to Flow."

28. Ibid., p. 121.

29. Culver, telephone interview with the author, October 22, 2008.

30. Bill Shiner, in telephone interview with the author, March 19, 2018.

31. Culver, telephone interview with the author, October 22, 2008.

32. Jeff Hecht, "History of Gas Lasers Part 1—Continuous-Wave Gas Lasers," *Optics & Photonics News*, January 1, 2010, https://www.osa-opn.org/home/articles/volume_21/issue_1/features/history_of_gas_lasers,_part_1%E2%80%94continuous_wave_gas/ (accessed May 15, 2018).

33. Jeff Hecht, *Beam: The Race to Make the Laser* (New York: Oxford, 2005), pp. 219–20.

34. R. J. Keyes and T. M. Quist, "Recombination Radiation Emitted by Gallium Arsenide," *Proc. IRE* 50 (August 1962): 1822–23.

35. Jeff Hecht, "The Breakthrough Birth of the Diode Laser," *Optics & Photonics News*, July 1, 2007, https://www.osa-opn.org/home/articles/volume_18/issue_7/features/the_breakthrough_birth_of_the_diode_laser/ (accessed May 15, 2018).

36. C. Koester and C. J. Campbell, "The First Clinical Application of the Laser," *Lasers*

in *Ophthalmology: Basic, Diagnostic, and Surgical Aspects; A Review*, ed. F. Fankhauser and S. Kwasniewska (Monroe, NY: The Hague: Kugler, 2003), pp. 115–17.

37. Loory, "Incredible Laser," pp. 6–7, 25.

38. Ibid.

39. Ibid.

40. Robert Hess, private communication, "A Survey of Lasers at the Birth of Holography," *Journal of Physics: Conference Series* 415, conference 1 (2013).

41. Seidel, "From Glow to Flow," p. 122.

42. Snitzer, in interview by Bromberg.

43. Seidel, "From Glow to Flow," p. 125.

44. Shiner, interview with the author, May 18, 2012.

45. Richard J. Barber, *The Advanced Research Projects Agency, 1958–1974: A Study* (Washington, DC: Richard J. Barber Associates, 1975), p. 6–11.

46. Robert W. Duffner, *Airborne Laser: Bullets of Light* (New York: Plenum, 1997), p. 16.

47. Sidney G. Reed, Richard H. Van Atta, and Seymour J. Deitchman, *DARPA Technical Accomplishments: An Historical Review of Selected DARPA Projects*, vol. 1 (Alexandria, VA: Institute for Defense Analyses, February 1990), p 8-2.

48. Dennis Overbye, "Arthur R. Kantrowitz, Whose Wide-Ranging Research Had Many Applications, Is Dead at 95," *New York Times*, December 9, 2008, http://www.nytimes.com/2008/12/09/science/09kantrowitz.html (accessed February 28, 2018).

49. Ben Bova, *Star Peace: Assured Survival* (New York: Tor, 1986), pp. 35–36.

50. Overbye, "Arthur R. Kantrowitz."

51. Howard Schlossberg, in interview with the author, March 7, 2018.

52. Arthur Kantrowitz, Oral History Interview by Joan Bromberg on October 30, 1984, Niels Bohr Library & Archives, American Institute of Physics, College Park, MD, www.aip.org/history-programs/niels-bohr-library/oral-histories/31415 (accessed September 28, 2018).

53. C. Kumar N. Patel, in interview with the author, in Jeff Hecht, ed., *Laser Pioneers: Revised Edition* (Boston, Academic Press, 1992), pp. 195–97.

54. Kantrowitz, interview with Bromberg.

55. Ibid.

56. Edward Gerry, in telephone interview with the author, October 6, 2008.

57. Gerry, telephone interview with the author.

58. Schlossberg, telephone interview with the author.

59. Gerry, telephone interview with the author.

60. Duffner, *Airborne Laser*, p. 17.

61. Gerry, telephone interview with the author.

62. "Bibliography," Ben Bova Online, last updated April 2018, http://benbova.com/bibliography/ (accessed May 16, 2018).

63. Bova, *Star Peace*, pp. 44–45.

64. Guest of honor speech at Boskone 1977, author attended, available online at History

of Boskone, http://www.nesfa.org/boskone-history/boskone-history.html (accessed May 16, 2018).

65. Jeff Hecht, *Laser Pioneers* (Boston: Academic Press, 1991), p. 41.

66. Ben Bova, "Chapter Seven," in *The Amazing Laser*, in *Out of the Sun* (New York: Tor, 1984); Frank Horrigan et al., "High-Power Gas Laser Research," *Final Technical Report May 31, 1967– April 30, 1968*, DTIC document AD0676226 (Fort Belvoir, VA: Defense Technical Information Center, July 1968) http://www.dtic.mil/docs/citations/AD0676226 (accessed February 28, 2018).

67. Seidel, "From Glow to Flow," p. 140.

68. Duffner, *Airborne Laser*, p. 24.

69. Schlossberg, telephone interview with the author.

70. Barber, *Advanced Research Projects Agency*, pp. viii—35.

71. Duffner, *Airborne Laser*, pp. 18–22.

72. N. G. Basov, in interview by Arthur Guenther, September 14, 1984, Niels Bohr Library & Archives, American Institute of Physics, College Park, MD, https://www.aip.org/history-programs/niels-bohr-library/oral-histories/4495 (accessed March 1, 2018); Wikipedia, s.v. "Nikolay Basov," last edited August 21, 2018, https://en.wikipedia.org/wiki/Nikolay_Basov (accessed August 30, 2018).

73. Duffner, *Airborne Laser*, pp. 20–25.

74. Jack Daugherty, in telephone interview with the author, April 3, 2018.

75. N. G. Basov and A. N. Oraevskii, "Attainment of Negative Temperatures by Heating and Cooling a System," *Soc. Phys. JETP* 17 (November 1963): 1171–74; V. K. Konyukhov and A. M. Prokhorov, "Population Inversion in Adiabatic Expansion of a Gas Mixture," *Journal of Experimental and Theoretical Physics Letters* 3 (June 1, 1966): 286–88.

76. Edward T. Gerry, "Gasdynamic Lasers," *IEEE Spectrum*, November 1970, pp. 51–58.

77. Schlossberg, telephone interview with the author.

78. Gerry, telephone interview with the author.

79. Wikipedia, s.v. "Almaz," last edited August 11, 2018, https://en.wikipedia.org/wiki/Almaz (accessed June 17, 2018).

80. Anatoly Zak, "The Soviet Laser Space Pistol, Revisited," *Popular Mechanics*, June 14, 2018, https://www.popularmechanics.com/space/satellites/a21527129/the-soviet-laser-space-pistol-revealed/ (accessed June 17, 2018).

81. Duffner, *Airborne Laser*, pp. 28–29.

82. Ibid., pp. 30–41.

83. Hecht, *Beam Weapons*, p. 284.

84. Duffner, *Airborne Laser*, pp. 41–42.

85. M. J. Soileau, in telephone interview with the author, March 19, 2018.

86. Ibid.

87. Ibid.

88. Richard Smith, in telephone interview with the author, July 5, 1994.

89. Joung Cook, "High-Energy Laser Weapons Since the Early 1960s," *Optical Engineering* 52, no. 3 (February 2013).

90. Ibid.

91. Ibid.

92. Gerry, "Gasdynamic Lasers."

93. J. V. V. Kasper and G. C. Pimentel, "HCl Chemical Laser," Physical Review Letters 14, no. 10 (1965): 352–54.

94. D. J. Spencer, H. Mirels, and T. A. Jacobs, "Initial Performance of a CW Chemical Laser," *Opto-Electronics* 2 (1970): 155–60.

95. Philip J. Klass, "Special Report: Laser Thermal Weapons, Power Boost Key to Feasibility," *Aviation Week & Space Technology*, August 21, 1972, pp. 32–40.

96. Spencer, Mirels, Jacobs, "Initial Performance."

97. Gerry, telephone interview with the author.

98. "Northrop Grumman Laser 'Firsts,'" Northrop Grumman, 2018, http://www.northropgrumman.com/Capabilities/LaserFirsts/Pages/default.aspx (accessed March 7, 2018).

99. Klass, "Special Report: Laser Thermal Weapons.""

100. *Aviation Week & Space Technology*, September 8, 1975, p. 53.

101. Philip J. Klass, "Laser Destroys Missile in Test," *Aviation Week & Space Technology*, August 7, 1978, pp. 14–16.

102. Hecht, *Beam Weapons*, p. 285.

103. "Northrop Grumman Laser 'Firsts.'"

104. "Laser Weaponry Technology Advances," *Aviation Week & Space Technology*, May 25, 1981, pp. 65–71.

105. William J. McCarthy, *Directed Energy and Fleet Defense: Implications for Naval Warfare* (Maxwell Air Force Base, AL: Center for Strategy and Technology, Air University, May 1980), p. 18 (cites references which are no longer posted on the web at the cited locations; some may be found at GlobalSecurity.org).

106. "Mid-Infrared Advanced Chemical Laser (MIRACL)," Global Security.org, last modified July 21, 2011, https://www.globalsecurity.org/space/systems/miracl.htm (accessed March 7, 2018) (republication of government document).

107. "James J. Ewing: Excimer Lasers," interviewed by C. Breck Hitz, in Hecht, *Laser Pioneers*: James J. Ewing, Excimer Lasers, pp. 243–56.

108. Hecht, *Laser Pioneers*, pp. 243–56.

109. Ibid., p. 247.

110. Ibid., pp. 249–50.

111. John Murray, "Lasers for Fusion Research," in *OSA Century of Optics* (Washington, DC: Optical Society of America, 2016), pp. 177–74.

112. Jeff Hecht, "Laser Isotope Enrichment," in *OSA Century of Optics* (Washington, DC: Optical Society of America, 2016), pp. 161–65.

113. Duffner, *Airborne Laser*, p. 64.

114. Ibid., p. 65.

115. Ibid., p. 64.

116. Hans Mark, "The Airborne Laser from Theory to Reality: An Insider's Account," *Defense Horizons*, April 2002.

117. Duffner, *Airborne Laser*, pp. 15–18.

118. Ibid., pp. 19–20.

119. "Major General Donald L. Lamberson," Biography, United States Air Force, April 28, 2009, https://www.af.mil/About-Us/Biographies/Display/Article/106449/major-general-donald-l-lamberson/ (accessed March 1, 2018).

120. Duffner, *Airborne Laser*.

121. Schlossberg, telephone interview with the author.

122. Hans Mark, "The Airborne Laser from Theory to Reality: An Insider's Account," *Defense Horizons*, April 2002.

123. Hecht, *Beam Weapons*, p. 286.

124. Duffner, *Airborne Laser*, p. 123–25.

125. Robert Duffner, *The Adaptive Optics Revolution* (Albuquerque: University of New Mexico Press, 2009), p. 224.

126. Duffner, *Adaptive Optics*, p. 212.

127. Ibid., p. 237 (photo of clean room).

128. Ibid., p. 206.

129. Ibid., pp. 234–35.

130. Ibid., pp. 235–37.

131. Ibid., pp. 235–36.

132. Ibid., pp. 237–39.

133. Ibid., pp. 248–53.

134. "Laser Fails to Destroy Missile," *Aviation Week & Space Technology*, June 8, 1981, p. 63.

135. Duffner, *Airborne Laser*, pp. 248–53.

136. Ibid., p. 279.

137. Ibid., pp. 286–89.

138. Julius Feinleib, "Toward Adaptive Optics," *Laser Focus* 10, no. 12 (December 1974): 44–70.

139. Harold "Hal" Walker, in telephone interviews with the author July 18, 2017 and April 18, 2018; "Hildreth 'Hal' Walker Jr.," Historical Inventors, Lemulson-MIT, https://lemelson.mit.edu/resources/hildreth-%E2%80%9Chal%E2%80%9D-walker-jr (accessed May 9, 2018).

140. Rod Waters, *Maiman's Invention of the Laser* (Amazon Digital Services, 2014), chap. 6.

141. Walker, telephone interviews with the author.

142. Duffner, *Airborne Laser*, pp. 303–305.

143. Ibid., pp. 306–10.

144. "Boeing NKC-135A Stratotanker (Airborne Laser Lab)," National Museum of the US Air Force, January 4, 2012, https://web.archive.org/web/20150722020529/http://www

.nationalmuseum.af.mil/factsheets/factsheet.asp?id=787 (accessed March 12, 2018). Photo of interior at "Boeing NKC-135A Cockpit," National Museum of the US Air Force, http://www.nationalmuseum.af.mil/Upcoming/Photos/igphoto/2000544279/ (accessed August 31, 2018).

CHAPTER 4: SPACE LASERS ON THE HIGH FRONTIER

1. Wikipedia, s.v. "Global Positioning System," last edited September 27, 2018, https://en.wikipedia.org/wiki/Global_Positioning_System (accessed September 28, 2018).

2. Bhupendra Jasani and Stockholm International Peace Research Institute, *Outer Space: A New Dimension of the Arms Race* (London: Taylor & Francis, 1982).

3. Neil Sheehan, *A Fiery Peace in a Cold War* (New York: Random House, 2007).

4. Donald R. Baucom, *The Origins of SDI: 1944–1983* (Lawrence: University Press of Kansas, 1992).

5. Ibid., p. 70.

6. Ibid., pp. 96–97.

7. Stockholm International Peace Research Institute, *The Arms Race and Arms Control* (London: Taylor and Francis, 1982), pp. 92–93.

8. Frances Fitzgerald, *Way Out There in the Blue* (New York: Simon & Schuster, 2000), pp. 93–97.

9. Philip J. Klass, "Laser Destroys Missile in Test," *Aviation Week & Space Technology*, August 7, 1978, pp. 14–16.

10. Michael J. Neufeld, "'Space Superiority'" Wernher von Braun's Campaign for a Nuclear-Armed Space Station, 1946–1956," *Space Policy* 22 (2006): 52–62.

11. Gerard K. O'Neill, "The Colonization of Space," *Physics Today* 27, no. 9 (1974): 32, https://doi.org/10.1063/1.3128863 (accessed March 13, 2018).

12. Ibid.

13. Ibid.

14. Elizabeth Howell, "Lagrange Points: Parking Places in Space," Space.com, August 21, 2017, https://www.space.com/30302-lagrange-points.html (accessed May 19, 2018).

15. "About the National Space Society," National Space Society, 2018, http://www.nss.org/about/ (accessed March 20, 2018).

16. Alvin Toffler, *Future Shock* (New York: Random House, 1970).

17. John Noble Wilford, "The Industrialization of Space: Why Industry Is Worried," *New York Times*, March 22, 1981, https://www.nytimes.com/1981/03/22/business/the-industrialilzation-of-space-why-business-is-wary-construction.html (accessed May 19, 2018).

18. Peter E. Glaser, "Earth Benefits of Solar Power Satellites," *Space Solar Power Review* 1, nos. 1 (January 1980): 9–38.

19. Gerard K. O'Neill, "The Low (Profile) Road to Space Manufacturing," *Astronautics and Aeronautics* 16, no. 3 (March 1978): 24–32.

20. Edward Gerry, in telephone interview with the author, April 11, 2018; "Member Spotlight: Peter Clark," Caltech Associates, 2018, http://associates.caltech.edu/member-spotlight-peter-clark (accessed May 20, 2018).

21. Sidney G. Reed, Richard H. Van Atta, and Seymour J. Deitchman, *DARPA Technical Accomplishments: An Historical Review of Selected DARPA Projects*, vol. 1 paper P-2192 (Alexandria, VA: Institute for Defense Analyses, February 1990), p. 8-7.

22. N. G. Basov, V. A. Danilychev, and Yu. M. Popov, "Stimulated Emission in the Vacuum Ultraviolet Region," *Soviet Journal of Quantum Electronics* 1, no. 1 (1971): 18–22, http://iopscience.iop.org/article/10.1070/QE1971v001n01ABEH003011/meta (accessed March 15, 2018).

23. James J. Ewing, interview by C. Breck Hitz, in Hecht, *Laser Pioneers*.

24. Jeff Hecht, "The History of the X-Ray Laser," *Optics & Photonics News* 19 (May 2008), https://www.osa-opn.org/home/articles/volume_19/issue_5/features/the_history_of_the_x-ray_laser/ (accessed May 21, 2018).

25. J. G. Kepros et al., "Experimental Evidence of an X-Ray Laser," *Proceedings of the National Academy of Sciences USA* 69 (1972): 1744–45.

26. Irving J. Bigio, email to author, July 28, 2008.

27. Ronald W. Waynant and Raymond C. Elton, "Review of Short-Wavelength Laser Research," *Proceedings of the IEEE* 64, no. 7 (July 1976): 1059–92.

28. Hecht, "The History of the X-Ray Laser."

29. John H. J. Madey, interview by C. Breck Hitz, "The Free-Electron Laser," in Hecht, *Laser Pioneers*, pp. 257–68.

30. L. Elias, W. Fairbank, J. Madey, H. Schwettman, and T. Smith, "Observation of Stimulated Emission of Radiation by Relativistic Electrons in a Spatially Periodic Transverse Magnetic Field," *Physical Review Letters* 36, no. 717 (March 29, 1976), https://journals.aps.org/prl/abstract/10.1103/PhysRevLett.36.717 (accessed May 21, 2018).

31. Sidney G. Reed, Richard H. Van Atta, and Seymour J. Deitchman, *DARPA Technical Accomplishments: An Historical Review of Selected DARPA Projects*, vol. 3, paper P-2538 (Alexandria, VA: Institute for Defense Analyses, July 1991), p. II-14–16.

32. Baucom, *Origins of SDI*, p. 109.

33. Louis Marquet, in telephone interview with the author, April 2, 2018.

34. Baucom, *Origins of SDI*.

35. "Laser Applications in Space Emphasized," *Aviation Week & Space Technology*, July 28, 1980, pp. 62–64.

36. Ibid.

37. Ibid.

38. Doris Hamill, in telephone interview with the author, June 7, 2018.

39. Ibid.

40. "Laser Applications in Space Emphasized," pp. 62–64.

41. Ibid.

42. Jack Daugherty, in telephone interview with the author, May 21, 2018.

43. Keith A. Truesdell, Charles A. Helms, and Gordon D. Hager, "History of Chemical Oxygen-Iodine Laser (COIL) Development in the USA," *Proceedings of the Society of PhotoOptical Instrumentation Engineers* 2502, *Gas Flow and Chemical Lasers: Tenth International Symposium* (March 31, 1995); doi: 10.1117/12.204917 (accessed May 21, 2018).

44. "Laser Weaponry Technology Advances," *Aviation Week*, May 25, 1981, pp. 65–69.

45. Baucom, *Origins of SDI*, pp. 118–20.

46. Ibid., pp. 120–21.

47. Ibid.

48. Ibid., p. 121.

49. Quoted in ibid., pp. 121–22.

50. Richard L. Garwin, "Boost-Phase Intercept: A Better Alternative," *Arms Control Today*, September 2000, https://www.armscontrol.org/act/2000_09/bpisept00 (accessed May 21, 2018).

51. Jeffrey T. Richelson, ed., *Space-Based Early Warning: From MIDAS to DSP to SBIRS* (Washington, DC: National Security Archive, electronic briefing book no. 235, November 9, 2007), https://nsarchive2.gwu.edu//NSAEBB/NSAEBB235/20130108.html (accessed May 22, 2018).

52. Baucom, *Origins of SDI*, p. 119.

53. Margalit Fox, "Malcolm Wallop, Senator from Wyoming, Dies at 78," *New York Times*, September 15, 2011, http://www.nytimes.com/2011/09/16/us/malcolm-wallop-ex-senator-of-wyoming-dies-at-78.html (accessed March 18, 2018).

54. Ibid.

55. Wikipedia, s.v. "Angelo Codevilla," last edited July 27, 2018, https://en.wikipedia.org/wiki/Angelo_Codevilla (accessed September 27, 2018).

56. Baucom, *Origins of SDI*, pp. 123–24.

57. Ibid., pp. 124–25.

58. Malcolm Wallop, "Opportunities and Imperatives of Ballistic Missile Defense," Strategic *Review* (Fall 1979): 13–21.

59. Ibid.

60. Baucom, *Origins of SDI*, pp. 125–26.

61. Ibid., p. 126.

62. Ibid.

63. "Defense Dept. Experts Confirm Efficacy of Space-Based Lasers," *Aviation Week & Space Technology*, July 28, 1980, pp. 65–66.

64. Baucom, *Origins of SDI*, p. 127.

65. Ibid.

66. Ibid., pp. 130–31.

67. *Report to the Congress on Space Laser Weapons* (Washington, DC: Office of the Under Secretary of Defense for Research and Engineering, February 26, 1981), declassified July 23,

2014, http://www.esd.whs.mil/Portals/54/Documents/FOID/Reading%20Room/Special_Collections/12-M-06360001.pdf (accessed May 22, 2018).

68. Ibid., pp. 2-2–2-3.

69. Ibid., p. 2-6.

70. Ibid., p. 2-7.

71. Ibid., p. 6-6.

72. Jeff Hecht, *Beam Weapons: The Next Arms Race* (New York: Plenum, 1984), p. 353.

73. Wikipedia, s.v. "Daniel O. Graham," last edited January 20, 2018, https://en.wikipedia.org/wiki/Daniel_O._Graham (accessed March 20, 2018).

74. Daniel O. Graham, *High Frontier: A New National Strategy* (Washington, DC: High Frontier, 1982), p. 8.

75. Graham, *High Frontier*, pp. 8–9.

76. George Chapline and Lowell Wood, "X-Ray Lasers," *Physics Today* 28, no. 6 (1975): 40; doi: 10.1063/1.3069004; https://doi.org/10.1063/1.3069004 (accessed September 4, 2018).

77. Ibid.

78. George Chapline, in interview with the author, February 7, 2008.

79. George Chapline, "Bootstrap Theory and Certain Properties of the Hadron Axial Vector Current" (PhD diss., California Institute of Technology, 1967), https://thesis.library.caltech.edu/4895/ (accessed September 4, 2018).

80. Chapline, telephone interview with the author.

81. "The Fireball," Atomic Archive, 2015, http://www.atomicarchive.com/Effects/effects8.shtml (accessed March 22, 2018).

82. P. V. Zarubin, "Academician Basov, High-Power Lasers, and the Antimissile Defense Problem," *Quantum Electronics* 32, no. 12 (2002): 1048–64.

83. Chapline, telephone interview with the author.

84. Ibid.

85. Ibid.

86. Anne M. Stark, "30 Years and Counting, the X-Ray Laser Lives On," Lawrence Livermore National Laboratory, April 14, 2015, https://www.llnl.gov/news/30-years-and-counting-x-ray-laser-lives (accessed March 22, 2018).

87. Clarence A. Robinson Jr., "Advance Made on High-Energy Laser," *Aviation Week & Space Technology*, February 23, 1981, pp. 25–27.

88. Ibid.

89. Hecht, *Beam Weapons*, pp. 126–31.

90. William Broad, *Teller's War* (New York: Simon & Schuster, 1992), p. 92.

91. Frances Fitzgerald, *Way Out There in the Blue* (New York: Simon & Schuster, 2000), p. 129.

92. "Review and Outlook 1975," *Laser Focus*, January 1975, pp. 10–28.

93. "Review and Outlook 1981," *Laser Focus*, January 1981, pp. 38–59.

94. Bill Shiner, in interviews with the author, May 18, 2012 and March 19, 2018.
95. Hal Walker, in telephone interview with the author, April 18, 2018.

CHAPTER 5: THE "STAR WARS" WARS

1. Ronald Reagan, "Address to the Nation on Defense and National Security" (speech, Oval Office, White House, Washington, DC, March 23, 1983), https://www.reaganlibrary.gov/sites/default/files/archives/speeches/1983/32383d.htm (accessed March 23, 2018).
2. Ed Gerry, in telephone interview with the author, April 11, 2018.
3. *Murder in the Air*, directed by Lewis Seiler, Warner Brothers, 1940.
4. Gerold Yonas, *Death Rays and Delusions* (Albuquerque, NM: Peter Publishing, 2017), p. 61.
5. "First Shuttle Launch," NASA, April 13, 2013 https://www.nasa.gov/multimedia/imagegallery/image_feature_2488.html (accessed May 22, 2018).
6. Frances Fitzgerald, *Way Out There in the Blue* (New York: Simon & Schuster, 2000), pp. 131–35.
7. Ibid., pp. 129–31.
8. Ibid., pp. 128.
9. Peter Goodchild, *Edward Teller: The Real Dr. Strangelove* (Cambridge, MA: Harvard, 2005).
10. Edward Teller with Judith Shooolery, *Memoirs* (Cambridge, MA: Perseus Publishing, 2001).
11. Dan O'Neill, *The Firecracker Boys* (New York: St. Martin's, 1994).
12. O'Neill, *Firecracker Boys*.
13. Fitzgerald, *Way Out There in the Blue*, p. 129.
14. John Nuckolls et al., "Laser Compression of Matter to Super High Densities: Thermonuclear Applications," *Nature* 239 (September 15, 1972): 139–42.
15. David Kramer, "NIF May Never Ignite, DoE Admits," *Physics Today*, June 2017, https://physicstoday.scitation.org/do/10.1063/PT.5.1076/full/ (accessed May 23, 2018).
16. *Memoirs*, p. 524.
17. William J. Broad, *Star Warriors* (New York: Simon & Schuster, 1985).
18. William J. Broad, *Teller's War* (New York: Simon & Schuster, 1992), p. 187.
19. Jeff Hecht, "Will We Catch a Falling Star?" *New Scientist*, September 7, 1991, https://www.newscientist.com/article/mg13117854-700/ (accessed May 23, 2018).
20. Howard Gluckman, "L5 News: Space, the Crucial Frontier—Citizens Advisory Council on National Space Policy," L5 News, April 1981, in National Space Society, http://space.nss.org/l5-news-space-the-crucial-frontier-citizens-advisory-council-on-national-space-policy/ (accessed September 27, 2018).
21. Jerry E. Pournelle, "*SPACE: The Crucial Frontier*" (Tucson, AZ: Citizens Advisory

Council on National Space Policy," Spring 1981), http://www.nss.org/settlement/L5news/L5news/CrucialFrontier1981.pdf (accessed March 24, 2018).

22. Ibid., p iii.

23. Wikipedia, s.v. "Citizens' Advisory Council on National Space Policy," last edited January 7, 2018, https://en.wikipedia.org/wiki/Citizens%27_Advisory_Council_on_National_Space_Policy (accessed March 24, 2018).

24. Ibid.

25. Baucom, *Origins of SDI*, p. 155.

26. Fitzgerald, *Way Out There in the Blue*, pp. 144–45.

27. Ibid., p. 145.

28. Sharon Watkins Lang, "Where Do We Get 'Star Wars,'" *Eagle*, March 2007, https://web.archive.org/web/20090227050446/http://www.smdc.army.mil/2008/Historical/Eagle/WheredowegetStarWars.pdf (accessed September 4, 2018).

29. Ryan Teague Beckwith, "George Lucas Wrote 'Star Wars'" as a Liberal Warning: Then Conservatives Struck Back," *Time*, October 10, 2017, http://time.com/4975813/star-wars-politics-watergate-george-lucas/ (accessed May 23, 2018).

30. Sharon Weinberger, *The Imagineers of War* (New York: Knopf, 2017), pp. 43–45.

31. "Washington Report," *Aviation Week & Space Technology*, April 23, 1984, p. 17.

32. Fitzgerald, *Way Out There in the Blue*, pp. 244–45.

33. Louis Marquet, in telephone interview with the author, April 2, 2018.

34. Ibid.

35. Ibid.

36. Fitzgerald, *Way Out There in the Blue*, p. 245.

37. Edward Teller, in letter to the author, January 10, 1984.

38. Author's notes from "Symposium on Lasers and Particle Beams for Fusion and Strategic Defense" (Rochester, NY: University of Rochester, April 17–19, 1985).

39. Ibid.

40. Jeff Hecht, "The History of the X-Ray Laser," *Optics & Photonics News*, May 2008, https://www.osa-opn.org/home/articles/volume_19/issue_5/features/the_history_of_the_x-ray_laser/ (accessed September 5, 2018).

41. Author's notes from "Symposium on Lasers and Particle Beams."

42. Yonas, *Death Rays and Delusions*, p. 101.

43. Ibid., pp. 102–103.

44. Ibid., p. 131.

45. Ibid., p. 132.

46. Ibid.

47. Broad, *Star Warriors*.

48. Ibid.

49. Ibid.

50. Ibid.

51. Yonas, *Death Rays and Delusions*, p. 186, citing US State Department Memorandum of Conversations, Reykjavik, Iceland, October 11–12, 1986.

52. Fitzgerald, *Way Out There in the Blue*, pp. 355–58.

53. Yonas, *Death Rays and Delusions*, pp. 183–89.

54. Gerold Yonas and Jill Gibson, "It's Laboratory or Goodbye," *STEPS Science Technology, Engineering, and Policy Studies*, no. 3 (February 18, 2006): 12–23, http://www.potomacinstitute.org/steps/featured-articles/65-it-s-laboratory-or-goodbye (accessed June 17, 2018).

55. Konstantin Lantratov, "The 'Star Wars' That Never Happened: The True Story of the Soviet Union's Polyus (Skif-DM) Space-Based Laser Battle Stations," *Quest: The History of Spaceflight Quarterly* 14, no. 1 (2007): 5–14; no. 2 (2007): 5–18.

56. Ibid.

57. Vassili Petrovitch, "Buran: Reusable Soviet Space Shuttle," Buran-Energia.com, http://www.buran-energia.com/bourane-buran/bourane-desc.php (accessed June 17, 2018).

58. Edward Reiss, *The Strategic Defense Initiative* (Cambridge, UK: Cambridge University Press, 1992), p. 79.

59. Broad, *Teller's War*, p. 160.

60. Nigel Hey, *The Star Wars Enigma* (Lincoln, NE: Potomac Books, 2006), p. 145.

61. "Teller's Telltale Letters," *Bulletin of the Atomic Scientists*, November, 1988, pp. 4–5.

62. Broad, *Teller's War*, pp. 162–64.

63. Ibid., p. 163.

64. Ibid., pp. 163–64.

65. Ibid., p. 169.

66. Ibid., p. 177.

67. Ibid., p. 187.

68. Ibid., p. 188.

69. R. Jeffrey Smith, "Experts Cast Doubts on X-Ray Laser," *Science*, November 8, 1985, p. 646.

70. Broad, *Teller's War*, pp. 204–205.

71. Ibid., pp. 200–201.

72. "Peter Hagelstein," MIT Directory, http://www.rle.mit.edu/people/directory/peter-hagelstein/ (accessed May 26, 2018).

73. Broad, *Teller's War*, p. 207.

74. Deborah Blum, "Weird Science: Livermore's X-Ray Laser Flap," *Bulletin of the Atomic Scientists* (July–August 1988): 7–13.

75. Ibid.

76. Broad, *Teller's War*, p. 19.

77. Daniel S. Greenburg, *Science, Money, and Politics: Political Triumph and Ethical Erosion* (Chicago, IL: University of Chicago Press, 2003), p. 292.

78. Richard L. Gullickson, "Advances in Directed Energy Technology for Strategic

Defense," *Proceedings, International Conference on Lasers "88*, Lake Tahoe, NV, December 4–9, 1988, pp. 270–77.

79. Ibid.

80. Timothy J. Seaman and William Dolman, *The 1986 GBFEL-TIE Sample Survey on White Sands Missile Range, New Mexico: The NASA, Stallion, and Orogrande Alternatives*, AD-A212 838 (Washington, DC: US Army Corps of Engineers, 1988).

81. Gary R. Goldstein, "Free-Electron Lasers as Ground Based Weapons," *AIP Conference Proceedings* 178, no. 290 (December 1988); https://doi.org/10.1063/1.37821 (accessed September 5, 2018).

82. "SDI Free Electron Laser Faces Cuts in Power, Delay," *Aviation Week & Space Technology*, May 22, 1989, p. 22.

83. "SDI Organization to Slash Funding for Ground-Based Free Electron Laser," *Aviation Week & Space Technology*, October 8, 1990, p. 27.

84. Anthony J. Cordi et al., "Alpha High-Power Chemical Laser Program," *Proceedings of SPIE* 1871, *Intense Laser Beams and Applications* (June 6, 1993).

85. Richard Ackerman et al., "Alpha High-Power Chemical Laser Program," *Proceedings of SPIE* 2502, *Gas Flow and Chemical Lasers: Tenth International Symposium* (March 31, 1995): 358–64.

86. "Space-Based Laser," *Aviation Week & Space Technology*, November 23, 1987, cover and pp. 80–81.

87. Craig Covault, "SDI Considers Cluster Booster to Launch Zenith Star Spacecraft," *Aviation Week & Space Technology*, November 30, 1987, pp. 20–21.

88. Theresa M. Foley, "Martin Marietta Hosts Reagan SDI Visit," *Aviation Week & Space Technology*, November 30, 1987, pp. 21–22.

89. "News Briefs," *Aviation Week & Space Technology*, December 12, 1988, p. 40.

90. Harry R. Finley, *Zenith Star Space-Based Chemical Laser Experiment* (Report NSIAD-89-118; Washington, DC, Government Accountability Office, April 1989), p. 3.

91. Ackerman et al., "Alpha High Power Chemical Laser Program."

92. James A. Horkovich, "Directed Energy Weapons: Promise & Reality" (37th AIAA Plasmadynamics and Lasers Conference, American Institute of Aeronautics and Astronautics, San Francisco, CA, June 5–8, 2006), https://arc.aiaa.org/doi/10.2514/6.2006-3753 (accessed June 17, 2018).

93. Fitzgerald, *Way Out There in the Blue*, p. 480.

94. Ibid., p. 482; Broad, *Teller's War*, pp. 251–54.

95. Robert L. Park, *Voodoo Science: The Road from Foolishness to Fraud* (London: Oxford, 2000), p. 188.

96. "SDI Organization to Slash Funding."

97. Michael A. Dornheim, "Alpha Chemical Laser Tests Affirm Design of Space-Based Weapon," *Aviation Week and Space Technology*, July 1, 1991, p. 26.

98. Ibid.

99. Fitzgerald, *Way Out There in the Blue*, pp. 484–85.

100. "Bring SDI Down to Earth," *Aviation Week & Space Technology*, February 18, 1991, p. 7; James R. Asker, "SDI Proposes $41 billion System to Stop Up to 200 Warheads," *Aviation Week & Space Technology*, February 18, 1991, p. 28–29.

101. Samuel W. Bowlin, *Strategic Defense Initiative: Changing Design and Technological Uncertainties Create Significant Risk* (Washington, DC: US General Accounting Office, February 19, 1992).

102. Fitzgerald, *Way Out There in the Blue*, pp. 488–89.

103. Ashlee Vance, "How an F Student Became America's Most Prolific Inventor," Bloomberg Business, October 20, 2015, https://www.bloomberg.com/features/2015-americas-top-inventor-lowell-wood/ (accessed May 27, 2018).

104. Yonas, *Death Rays and Delusions*.

105. Gerold Yonas, email to the author, March 29, 2018.

CHAPTER 6: THE AIRBORNE LASER GETS OFF THE GROUND

1. Hans Mark, "The Airborne Laser from Theory to Reality, an Insider's Account," *Defense Horizons*, no. 12 (April 2002): 10.

2. Ibid.

3. Keith A. Truesdell, Charles A. Helms, and Gordon D. Hager, "History of Chemical Oxygen-Iodine Laser (COIL) Development in the USA," *Proceedings of the Society of Photo-Optical Instrumentation Engineers*, 2502, *Gas Flow and Chemical Lasers: Tenth International Symposium* (March 31, 1995), doi: 10.1117/12.204917.

4. Ibid.

5. W. E. McDermott, N. R. Pchelkin, D. J. Benard, and R. R. Bousek, "An Electronic Transition Chemical Laser," *Applied Physics Letters* 32, no. 469 (1978), https://aip.scitation.org/doi/abs/10.1063/1.90088 (accessed April 4, 2018).

6. Truesdell, Helms, and Hager, "History of Chemical Oxygen-Iodine Laser."

7. Airborne Laser System Program Office, Public Affairs, "A Brief History of the Airborne Laser" (fact sheet), February 27, 2003.

8. H. W. Babcock, "The Possibility of Compensating Astronomical Seeing," *Publications of the Astronomical Society of the Pacific* 65 (October 1953): 229–36.

9. Robert Duffner, *The Adaptive Optics Revolution: A History* (Albuquerque: University of New Mexico Press, 2009), pp. 31–32.

10. Julius Feinleib, "Toward Adaptive Optics," *Laser Focus* 10 & 12 (December 1974): 44, 69–70.

11. Duffner, pp. 134–42.

12. Louis Marquet, telephone interview with the author, April 2, 2018.

13. Philip J. Klass, "Adaptive Optics Evaluated as Laser Aid," *Aviation Week & Space Technology*, August 24, 1981, pp. 61–65.

14. "'Remote Censoring': DOD Blocks Symposium Papers," *Science News* 122 (September 4, 1982): 148–49.

15. Janet Fender, in telephone interview with the author, May 16, 2018.

16. Katherine A. Finlay, *Hubble Space Bi-Stem Thermal Shield Analysis* (Cleveland, OH: NASA Glenn Research Center, September 1, 2004), https://ntrs.nasa.gov/archive/nasa/casi.ntrs.nasa.gov/20050186774.pdf (accessed June 15, 2018).

17. Fender, telephone interview with the author.

18. Janet S. Fender and Richard A. Carreras, "Demonstration of an Optically Phased Telescope Array," *Optical Engineering* 27 (September 1988): 706.

19. Duffner, *Adaptive Optics Revolution*, p. 205.

20. "Phillips Lab Studies Antimissile Laser," *Aviation Week & Space Technology*, December 7, 1992, pp. 51–53.

21. Ibid.

22. Ibid.

23. Stacey Evers, "Boeing, Rockwell Confront Airborne Laser Challenges," *Aviation Week & Space Technology*, May 30, 1994, p. 75.

24. William B. Scott, "Tests Support Airborne Laser as Viable Missile Killer," *Aviation Week & Space Technology*, December 19, 1994.

25. Ken Billman, in telephone interview with the author, April 17, 2015.

26. Truesdell, Helms, and Hager, "History of Chemical Oxygen-Iodine Laser."

27. Scott, "Tests Support Airborne Laser."

28. David A. Fulgham, "USAF Aims Laser at Antimissile Role," *Aviation Week & Space Technology*, August 14, 1995, pp. 24–25.

29. David A. Fulgham, "USAF Sees New Roles for Airborne Laser," *Aviation Week & Space Technology*, October 7, 1996, pp. 26–27.

30. Wikipedia, s.v. "Space Shuttle," last edited August 31, 2018, https://en.wikipedia.org/wiki/Space_Shuttle (accessed September 6, 2018).

31. Michael E. Davey and Frederick Martin, *The Airborne Laser Anti-Missile Program* (Washington, DC: Congressional Research Service, February 18, 2000), p. CRS-2–3.

32. Ibid.

33. Ibid., pp. CRS-22–23.

34. Kenneth Billman, in telephone interview with the author, April 17, 2015.

35. Ibid.

36. Wikipedia, s.v. "Itek," last edited June 30, 2018, https://en.wikipedia.org/wiki/Itek (accessed April 10, 2018).

37. Billman, telephone interview with the author.

38. Hassaun Jones-Bey, "Evolving JDS Uniphase to Acquire OCLI," *Laser Focus World*,

November 15, 1999, https://www.laserfocusworld.com/articles/1999/11/evolving-jds-uniphase-to-acquire-ocli.html (accessed April 10, 2018).

39. Billman, telephone interview with the author.

40. Davey and Martin, "*Airborne Laser Anti-Missile Program*," pp. CRS-25–30.

41. Ibid.

42. Gen. Ellen Pawlikowski, in telephone interview with the author, April 25, 2018.

43. Ibid.

44. Ibid.

45. Ibid.

46. Ibid.

47. Ibid.

48. Ibid.

49. Ibid.

50. *Cost Increases in the Airborne Laser Program*, GAO-04-643R (Washington, DC: US General Accounting Office, May 17, 2004), p. 2.

51. Ibid.

52. Carlin Leslie, "Women's AF History Expands with New Four-Star," US Air Force News, June 2, 2015, https://www.af.mil/News/Article-Display/Article/590267/womens-af-history-expands-with-new-four-star/ (accessed September 6, 2018); "General Ellen M. Pawlikowski," US Air Force, last updated June 2015, http://www.af.mil/About-Us/Biographies/Display/Article/104867/lieutenant-general-ellen-m-pawlikowski/ (accessed May 25, 2018).

53. "General Ellen M. Pawlikowski," US Air Force.

54. Pawlikowski, telephone interview with the author.

55. Ibid.

56. Michael Fabey, "Light Show," *Aviation Week*, June 18, 2007, pp. 172–73.

57. Ken Billman, in telephone interview with the author, May 1, 2018.

58. Pawlikowski, telephone interview with the author.

59. Billman, telephone interview with the author, May 1, 2018.

60. Ibid.

61. Arthur C. Clarke, "'Hazards of Prophecy: The Failure of Imagination,'" in *Profiles of the Future: An Enquiry into the Limits of the Possible*, rev. ed. (1962; New York: Harper, 1973), pp. 14, 21, 36.

62. Ken Billman, email to the author, June 5, 2018.

63. Christopher Bolkcom and Steven Hildreth, *Airborne Laser (ABL): Issues for Congress* (Washington, DC: Congressional Research Service, July 9, 2007), pp. CRS-9–10.

64. Ibid.

65. Ibid.

66. Ibid.

67. Ibid.

68. Ibid.

69. Paul Marks, "Airborne Laser Lets Rip on First Target," *New Scientist* 13 (December 10, 2008), https://www.newscientist.com/article/mg20026866.200-airborne-laser-lets-rip-on-first-target/ (accessed April 12, 2018).

70. Ibid.

71. Dominic Gates, "Boeing Hit Harder than Rivals by Defense Cuts," *Seattle Times*, April 7, 2009, https://web.archive.org/web/20090410052937/http://seattletimes.nwsource.com/html/localnews/2008997361_defensecuts07.html (accessed April 12, 2018).

72. *Hearings Before a Subcommittee of the Comm. on Appropriations*, 111th Cong. (2009) (statement of Robert Gates, Secretary of Defense), https://www.gpo.gov/fdsys/pkg/CHRG-111hhrg56285/pdf/CHRG-111hhrg56285.pdf (accessed September 29, 2018); quoted by Philip Coyle, in email to author, September 26, 2017.

73. Ibid.

74. Jeff Hecht, "Airborne Laser Still Firing," *Laser Focus World*, July 15, 2009, https://www.laserfocusworld.com/ore/en/articles/print/volume-16/issue-14/features/airborne-laser-still-firing.html (accessed April 17, 2018).

75. Amy Butler, "Ray of Light," *Aviation Week & Space Technology*, February 22, 2010, pp. 26–27.

76. Bolkcom and Hildreth, *Airborne Laser*.

77. Amy Butler, "Lights Out," *Aviation Week & Space Technology*, January 2, 2012, pp. 29–30.

78. Ed Gerry, telephone interview with author, April 11, 2018.

79. Fender, telephone interview with the author.

80. Pawlikowski, telephone interview with the author.

81. John Wallace, "Airborne Laser Test Bed Is Put to Rest in the Boneyard," *Laser Focus World*, February 17, 2012, https://www.laserfocusworld.com/articles/2012/02/airborne-laser-test-bed-is-put-to-rest-in-the-boneyard.html (accessed April 17, 2018).

82. Allen Nogee, "The Death of a Giant Laser," Strategies Unlimited, May 6, 2014, https://www.strategies-u.com/articles/2014/05/the-death-of-a-giant-laser.html (accessed April 12, 2018).

CHAPTER 7: BACK TO THE BATTLEFIELD AGAINST INSURGENTS

1. Josef Shwartz, Gerald Wilson, and Joel Avidor, "Tactical High Energy Laser," *SPIE Proceedings on Laser and Beam Control Technologies* 4632 (January 21, 2002), http://www.northropgrumman.com/Capabilities/ChemicalHighEnergyLaser/TacticalHighEnergyLaser/Documents/pageDocuments/SPIE_Manuscript_Tactical_high-.pdf (accessed April 13, 2018).

2. Ibid.

3. Sharon Watkins Lang, "SMCD History: Putting THEL on the Fast Track," Army Space and Missile Defense Command, May 4, 2017, https://www.army.mil/article/187148/smdc_history_putting_thel_on_the_fast_track (accessed April 13, 2018).

4. "US-Israel Strategic Cooperation: Tactical High-Energy Laser Program," Jewish Virtual Library, last updated January 2006, http://www.jewishvirtuallibrary.org/tactical-high-energy-laser-program (accessed April 14, 2018).

5. Lang, "SMCD History."

6. Shwartz, Wilson, and Avidor, "Tactical High Energy Laser."

7. Ibid.

8. "Mobile Tactical High-Energy Laser," GlobalSecurity.org, https://www.globalsecurity.org/space/systems/mthel.htm (accessed April 13, 2018).

9. Ibid.

10. Ibid.

11. Barbara Opall-Rome, "Israeli Military Resists Additional THEL Funding," *Defense News*, September 17–23, 2001, p. 38.

12. "Mobile Tactical High-Energy Laser."

13. Robert Wall, "Army Advances Tactical Lasers," *Aviation Week & Space Technology*, January 1, 2001, pp. 57, 60.

14. Jeff Hecht, "Advanced Tactical Laser Is Ready for Flight Tests," *Laser Focus World*, February 1, 2008, https://www.laserfocusworld.com/articles/2008/02/laser-weapons-advanced-tactical-laser-is-ready-for-flight-tests.html (accessed April 15, 2018).

15. Ibid.

16. Ibid.

17. "Fact Sheet: Advanced Tactical Laser," US Air Force Research Laboratory, August 2006.

18. Hecht, "Advanced Tactical Laser Is Ready."

19. John Wallace, "Boeing's Advanced Tactical Laser 'Defeats' Ground Target in Flight Test," *Laser Focus World*, September 2, 2009, https://www.laserfocusworld.com/articles/2009/09/boeings-advanced-tactical-laser-defeats-ground-target-in-flight-test.html (accessed April 15, 2018).

20. "Boeing Advanced Tactical Laser in Action," theworacle, October 1, 2009, YouTube video, 0:17, https://www.youtube.com/watch?v=qfmEUqmgsK4 (accessed April 15, 2018).

21. Marc Selinger and Chuck Cadena, "Boeing Advanced Tactical Laser Strikes Moving Target in Test," PR Newswire, Oct 13, 2009, http://boeing.mediaroom.com/2009-10-13-Boeing-Advanced-Tactical-Laser-Strikes-Moving-Target-in-Test (accessed April 15, 2018).

22. Mark Neice, in telephone interview with the author, April 17, 2018.

23. "Airborne Tactical Laser Feasibility for Gunship Operations Abstract," Air Force Scientific Advisory Board, 2008, http://www.scientificadvisoryboard.af.mil/Portals/73/documents/AFD-151215-017.pdf (accessed April 15, 2018).

24. Stuart Fox, "Pew. Airborne Military Laser Takes out Truck on Video," *Popular*

Science, October 1, 2009, https://www.popsci.com/MILITARY-AVIATION-AMP-SPACE/ARTICLE/2009-10/PEW-AIRBORNE-MILITARY-LASER-TAKES-OUT-TRUCK (accessed April 15, 2018).

25. "Airborne Tactical Laser Feasibility."

26. Neice, telephone interview with the author.

27. "Airborne Tactical Laser Feasibility."

28. They are Isamu Akasaki, Hiroshi Amano, and Shuji Nakamura, https://www.nobelprize.org/prizes/physics/2014/summary/ (accessed September 30, 2018).

29. *Defense Science Board Task Force on Directed Energy Weapons* (Washington, DC: Office of the Under Secretary of Defense for Acquisition, Technology and Logistics, December 2007), p. viii.

30. Ibid., p. xiv.

31. Ibid., p. 27.

32. "Navy Readies Free-Electron Laser Contracts," Defense Daily, January 5, 2009.

33. David Smalley, Office of Naval Research spokesman, in telephone interview with the author, April 11, 2013.

CHAPTER 8: LASER WEAPONS GO SOLID-STATE

1. *Report of the High Energy Laser Executive Review Panel: Department of Defense Laser Master Plan* (Washington, DC: Department of Defense, March 24, 2000), http://www.wslfweb.org/docs/MasterLaserPlan.pdf (accessed April 17, 2018).

2. C. A. Brau, "The Development of Very High Power Lasers," in *Proceedings of the International Conference on Lasers '88*, ed. R. C. Sze, F. J. Duarte, and the Society for Optical and Quantum Electronics (McLean, VA: STS Press, 1989), pp. 20–32.

3. *Report of the High Energy Laser.*

4. Brau, "Development of Very High Power Lasers."

5. *Report of the High Energy Laser.*

6. Ibid.

7. Ibid.

8. Ibid., table 1, p. 9.

9. Ibid.

10. *Form S-4: JDS Uniphase Corp.* (Washington, DC: US Securities and Exchange Commission, Morningstar Document Research, September 7, 2000), p. 30.

11. *Report of the High Energy Laser.*

12. Ibid.

13. Ibid.

14. Mark Neice, in telephone interview with the author, April 17, 2018.

15. Ibid.

16. Ed Pogue, director of HEL-JTO, in telephone interview with the author, July 8, 2004.
17. Ann Parker, "World's Most Powerful Solid-State Laser," *Science & Technology Review*, October 2002, https://str.llnl.gov/str/October02/Dane.html (accessed April 18, 2018).
18. Jeff Hecht, "Laser Weapons Go Solid-State," *Laser Focus World*, September 1, 2004, https://www.laserfocusworld.com/articles/print/volume-40/issue-9/features/back-to-basics-solid-state-lasers/laser-weapons-go-solid-state.html (accessed April 18, 2018).
19. Jeff Hecht, "Ray Guns Get Real," *IEEE Spectrum*, June 30 2009, https://spectrum.ieee.org/semiconductors/optoelectronics/ray-guns-get-real (accessed April 23, 2018).
20. Hecht, "Laser Weapons."
21. Pogue, telephone interview with the author.
22. Ibid.
23. "History," General Atomics, http://www.ga.com/history (accessed April 20, 2018).
24. George Dyson, *Project Orion: The True Story of the Atomic Spaceship* (New York: Henry Holt, 2003).
25. "History," General Atomics.
26. "Future Optics: Optics and Photonics Move Remotely Piloted Aircraft Forward, an Interview with Michael D. Perry," *Laser Focus World*, December 16, 2016, https://www.laserfocusworld.com/articles/print/volume-52/issue-12/columns/future-optics/future-optics-optics-and-photonics-move-remotely-piloted-aircraft-forward-an-interview-with-michael-d-perry.html (accessed April 20, 2018).
27. Michael D. Perry et al., "Laser Containing a Distributed Gain Medium," US Patent 7366211B2, filed September 5, 2006, https://patents.google.com/patent/US7366211B2/en (accessed April 20, 2018).
28. "Used 2003 Cadillac Escalade Features & Specs," Edmunds.com, https://www.edmunds.com/cadillac/escalade/2003/features-specs/ (accessed May 29, 2018).
29. "The IPG Story Is about Independence, Creating Advanced Technology, Having a Unique Business Model, and Rampant Success," IPG Photonics, http://www.ipgphotonics.com/en/whyIpg#[history] (accessed April 20, 2018).
30. Ibid.
31. P. Sprangle et al., "High-Power Fiber Lasers for Directed Energy Applications," *Naval Research Laboratory Review* (2008): 89–99, https://www.nrl.navy.mil/content_images/08FA3.pdf (accessed April 20, 2018).
32. National Research Council, *Review of Directed Energy Technology for Countering Rockets, Artillery, and Mortars (RAM)*, abbreviated version (Washington, DC: National Academies Press, 2008), pp. 1–2, http://nap.edu/12008 (accessed April 19, 2018).
33. Ibid.
34. Ibid.
35. Ibid.
36. Ibid., pp. 4–5.
37. "IPG Story."

38. National Research Council, *Review of Directed Energy*, pp. 4–5.

39. Gail Overton, "IPG Offers World's First 10-kW Single-Mode Production Laser," *Laser Focus World*, June 17, 2009, https://www.laserfocusworld.com/articles/2009/06/ipg-photonics-offers-worlds-first-10-kw-single-mode-production-laser.html (accessed May 29, 2018).

40. National Research Council, *Review of Directed Energy*, pp. 4–5.

41. John Wachs, Army Space and Missile Defense Command, Huntsville, AL, in telephone interview with the author, January 9, 2009.

42. Ibid.

43. Dan Wildt, in author's notes from press teleconference, March 18, 2009.

44. Hecht, "Ray Guns Get Real."

45. "Laureates 2010: Information Technology/Electronics Laureate, Joint High Power Solid-State Laser," *Aviation Week*, March 29/April 5, 2010, p. 70.

46. David Belforte, "What's New with 100-kW Fiber Lasers?" *Industrial Laser Solutions*, January 26, 2015, https://www.industrial-lasers.com/articles/print/volume-30/issue-1/departments/update/what-s-new-with-100kw-fiber-lasers.html (accessed May 29, 2018).

47. "Model Information: 2013 Honda Fit," Honda Owners, http://owners.honda.com/vehicles/information/2013/Fit/specs#mid^GE8G3DEXW (accessed April 23, 2018).

48. Neice, telephone interview with the author.

49. Author notes from Solid-State and Diode Laser Technology Review conference held by the Directed Energy Professional Society, June 29–July 2, 2009. (Technical digest was published.)

50. Ibid.

51. Ibid.

52. Ibid.

53. Ibid.

54. Eric Beidel, "All Systems Go: Navy's Laser Weapon Ready for Summer Deployment," US Office of Naval Research, April 7, 2014, https://www.onr.navy.mil/en/Media-Center/Press-Releases/2014/Laser-Weapon-Ready-For-Deployment (accessed April 24, 2018).

55. "Laser Weapon System (LaWS)," US Navy research, December 10, 2014, YouTube video, 1:25, http://youtu.be/D0DbgNju2wE (accessed April 24, 2018).

56. David Smalley, "Historic Leap: Navy Shipboard Laser Operates in Persian Gulf," US Office of Naval Research, December 10, 2014, https://www.onr.navy.mil/Media-Center/Press-Releases/2014/LaWS-shipboard-laser-uss-ponce (accessed April 24. 2018).

57. "Raytheon, Lockheed Share $23.8 Million for Laser Weapon," Optics.org, July 8, 2010, http://optics.org/news/1/2/2 (accessed April 23, 2018).

58. David Filgas et al., "Recent Results for the Raytheon RELI Program," *Proceedings SPIE* 8381, *Laser Technology for Defense and Security* 8, 83810W (May 7, 2012), https://www.spiedigitallibrary.org/conference-proceedings-of-spie/8381/83810W/Recent-results-for-the-Raytheon-RELI-program/10.1117/12.921055.full (accessed April 24, 2018).

59. Gregory D. Goodno et al., "Diffractive Coherent Combining of >kW Fibers," in

CLEO 2014, OSA Technical Digest (Optical Society of America, May 2014), paper STh4N.3, https://www.researchgate.net/publication/263785967 (accessed April 24, 2018).

60. Robert Afzal, Lockheed Martin, in telephone interview with the author, July 10, 2017.

CHAPTER 9: THE QUEST FOR THE ULTIMATE WEAPON

1. "Protocol on Blinding Laser Weapons," International Committee of the Red Cross, October 13, 1995, https://ihl-databases.icrc.org/applic/ihl/ihl.nsf/Treaty.xsp?action=openDocument&documentId=70D9427BB965B7CEC12563FB0061CFB2 (accessed May 29, 2018).

2. Janet Fender, in telephone interview with the author, May 16, 2018.

3. Ibid.

4. Ibid.

5. Ibid.

6. Ibid.

7. Janet Fender, in interview with the author, May 16, 2018.

8. *Austin Powers: International Man of Mystery*, directed by Jay Roach, New Line Cinema, 1997.

9. Jeff Hecht, "Laser Dazzlers Are Deployed," *Laser Focus World*, March 1, 2012, https://www.laserfocusworld.com/articles/print/volume-48/issue-03/world-news/laser-dazzlers-are-deployed.html (accessed May 29, 2018).

10. "Iron Beam: High-Power Mobile Laser Weapon System," Rafael Advanced Defense Systems, 2018, http://www.rafael.co.il/5688-763-en/Marketing.aspx (accessed April 25, 2018).

11. Oliver Hoffmann, "The High-Energy Laser: Weapon of the Future Already a Reality at Rheinmetall," Rheinmetall, https://www.rheinmetall.com/en/rheinmetall_ag/press/themen_im_fokus/zukunftswaffe_hel/index.php (accessed April 25, 2018).

12. "China Developing Portable Laser Weapons to Shoot Down Terrorist Drones to Enemy Missiles and Satellites," International Defense, Science and Technology, November 18, 2017, http://idstch.com/home5/international-defence-security-and-technology/military/land-230/china-developing-portable-laser-weapons-to-shoot-down-terrorist-drones-to-enemy-missiles-and-satellites/ (accessed April 25, 2018).

13. George Allison, "Dragonfire, the New British Laser Weapon," UK Defense Journal, June 5, 2017, https://ukdefencejournal.org.uk/dragonfire-new-british-laser-weapon/ (accessed April 25, 2018).

14. Tom O'Connor, "Russia's Military Has Laser Weapons That Can Take out Enemies in Less than a Second," *Newsweek*, March 12, 2018, http://www.newsweek.com/russia-military-laser-weapons-take-out-enemies-less-second-841091 (accessed April 25, 2018).

15. Spokesman for Public Affairs Department, Office of Naval Research, in telephone interview with the author, August 4, 2017.

16. Wikipedia, s.v. "USS Ponce (LPD-15)," last edited August 13, 2018, https://en.wikipedia.org/wiki/USS_Ponce_(LPD-15) (accessed April 25, 2018).

17. Spokesman for Public Affairs Department, telephone interview with the author.

18. Ibid.

19. James LaPorta, "Navy Orders Laser Weapon Systems from Lockheed Martin," United Press International, January 29, 2018, https://www.upi.com/Defense-News/2018/01/29/Navy-orders-laser-weapon-systems-from-Lockheed-Martin/6831517242381/ (accessed April 26, 2018).

20. Robert Afzal, speaking at Lockheed press teleconference, March 1, 2018.

21. Jeff Hecht, "Lockheed Martin to Develop Laser Weapon System for US Navy Destroyers," *IEEE Spectrum*, March 2, 2018, https://spectrum.ieee.org/tech-talk/aerospace/military/lockheed-martin-develops-helios-laser-weapon-for-us-navy (accessed April 26, 2018).

22. Afzal, speaking at Lockheed press teleconference.

23. "Directed Energy Directorate Laser Weapon Systems" (Air Force Research Laboratory, December 2016), http://www.kirtland.af.mil/Portals/52/documents/LaserSystems.pdf (accessed April 26, 2018).

24. Ibid.

25. Robert Afzal, telephone interview with the author, July 10, 2017.

26. Matthew Cox, "Marines Developing JLTV Air-Defense System Armed with Laser Weapon," Military.com, March 21, 2018, https://www.military.com/defensetech/2018/03/21/marines-developing-jltv-air-defense-system-armed-laser-weapon.html (accessed April 30, 2018).

27. Todd South, "New Pentagon Research Chief Is Working on Lasers, AI, Hypersonic Munitions, and more," *Military Times*, April 24, 2018, https://www.militarytimes.com/news/your-military/2018/04/24/new-pentagon-research-chief-is-working-on-lasers-ai-hypersonic-munitions-and-more/ (accessed April 27, 2018).

28. Leigh Giangreco, "Missile Defense Agency Solicits Industry for Low Power Laser Demonstration," Flight Global, September 2, 2016, https://www.flightglobal.com/news/articles/missile-defense-agency-solicits-industry-for-low-pow-428992/ (accessed April 27, 2018).

29. Michael Perry, in telephone interview with the author, April 27, 2018.

30. James Drew, "MDA Advances Missile-Hunting UAV Programs," *Aviation Week & Space Technology*, February 16, 2018.

31. Ibid.

32. Ibid.

33. Ibid.

34. Jeff Hecht, *Beam: The Race to Make the Laser* (New York: Oxford University Press, 2005), p. 41; Jeff Hecht, *Laser Pioneers* (Boston: Academic Press, 1991), p. 88.

35. Hecht, *Beam*, pp. 219–20.

36. William Krupke, in telephone interview, February 22, 2010.

37. Ibid.

38. Ibid.

39. William F. Krupke, "Diode Pumped Alkali Lasers (DPALs): A Review," *Progress in Quantum Electronics* 36, no. 1 (January 2012): 4–28.

40. In the laser world, semiconductor lasers usually are called "diodes" because the devices have two electrical terminals, making them electrical diodes. But diodes are laser geek speak that I try to avoid in this book.

41. Perry, telephone interview with the author.

42. Donald J. Kessler and Burton G. Cour-Palais, "Collision Frequency of Artificial Satellites: The Creation of a Debris Belt," *Journal of Geophysical Research: Space Physics*, 83, no. A6 (June 1, 1978): 2367–2646, https://agupubs.onlinelibrary.wiley.com/doi/abs/10.1029/JA083iA06p02637 (accessed April 30, 2018).

43. Rémi Soulard, Mark N Quinn, Toshiki Tajima, and Gérard Mourou, "ICAN: A Novel Laser Architecture for Space Debris Removal," *Acta Astronautica* 105, no. 1 (December 2014): 192–200, http://dx.doi.org/10.1016/j.actaastro.2014.09.004 (accessed April 30, 2018).

44. Alison Gibbings et al., "Potential of Laser-Induced Ablation for Future Space Applications," *Space Policy* 28 (2012): 149–53, http://dx.doi.org/10.1016/j.spacepol.2012.06.008 (accessed May 1, 2018).

45. Franklin B. Mead Jr., *Part 1: The Lightcraft Technology Demonstration Program* (Edwards, CA: Air Force Research Lab, Edwards AFB, November 2007), AFRL-RZ-ED-TR-2007-0078, http://www.dtic.mil/dtic/tr/fulltext/u2/a475937.pdf (accessed April 30, 2018).

46. Franklin B. Mead Jr. et al., *Advanced Propulsion Concepts: Project Outgrowth* (Edwards, CA: Air Force Rocket Propulsion Laboratory, AFRPL-TR-72-32, June 1972), pp. II-53–II-63, http://www.dtic.mil/dtic/tr/fulltext/u2/750554.pdf (accessed April 30, 2018).

47. Arthur Kantrowitz, "Propulsion to Orbit by Ground-Based Lasers," *Astronautics and Aeronautics* 10, no. 5 (May 1972): 74–76.

48. Omer F. Spurlock, *Performance Capability of Laser-Powered Launch Vehicles Using Vertical Ascent Trajectories* (Washington, DC: NASA Technical Memorandum, August 1974), https://ntrs.nasa.gov/archive/nasa/casi.ntrs.nasa.gov/19740023219.pdf (accessed May 1, 2018).

49. "LaserMotive Wins $900,000 from NASA in Space Elevator Games," NASA, November 9, 2009, https://www.nasa.gov/centers/dryden/status_reports/power_beam.html (accessed May 1, 2018).

50. Graham Warwick, "Power via Laser Beam Moves Closer to Reality," *Aviation Week & Space Technology*," April 24, 2018, https://aviationweek.com/connected-aerospace/power-laser-beam-moves-closer-reality? (accessed May 1, 2018).

51. Alison Power et al., "Potential of Laser-Induced Ablation for Future Space Applications," *Space Policy* 28 (2012): 149–53, http://dx.doi.org/10.1016/j.spacepol.2012.06.008 (accessed May 1, 2018).

52. Shelly Leachman, "California Scientists Propose System to Vaporize Asteroids That

Threaten Earth," University of California at Santa Barbara, February 14, 2013, http://www.news.ucsb.edu/2013/013465/california-scientists-propose-system-vaporize-asteroids-threaten-earth (accessed May 1, 2018).

53. "DE-STAR: Directed Energy Planetary Defense," UCSB Experimental Cosmology Group, http://www.deepspace.ucsb.edu/projects/directed-energy-planetary-defense (accessed May 1, 2018); Philip Lubkin, in telephone interview with the author, February 18, 2013.

54. Dennis Overbye, "Reaching for the Stars across 4.37 Light Years," *New York Times*, April 12, 2016, https://www.nytimes.com/2016/04/13/science/alpha-centauri-breakthrough-starshot-yuri-milner-stephen-hawking.html (accessed May 1, 2018).

55. Arati Prabhakar, in interview with the author, April 5, 2018.

56. Ibid.

57. Ellen Pawlikowski, in interview with the author, April 25, 2018.

58. Ibid.

59. James Horkovich, in interview with the author, May 2, 2018.

60. South, "New Pentagon Research Chief."

61. Barbara Opall-Rome, "Experts Tout Space-Based Sensors, Lasers for Missile Defense," *Defense News*, September 6, 2017, https://www.defensenews.com/smr/defense-news-conference/2017/09/06/us-experts-tout-space-based-sensors-lasers-for-active-defense/ (accessed April 27, 2018).

62. Ibid.

63. "Program of Record," *Acquisition Encyclopedia*, Defense Acquisition University, https://www.dau.mil/acquipedia/Pages/ArticleDetails.aspx?aid=2f2b8d1e-8822-4f88-9859-916ad81b597e (accessed May 3, 2018).

64. Richard Stimson, "Wrights' Perspective on the Role of Airplanes in War," The Wright Stories, http://wrightstories.com/wrights-perspective-on-the-role-of-airplanes-in-war/ (accessed May 3, 2018).

65. Ibid.

66. Ibid.

67. Ibid.

68. Ibid.

69. *Testimony Before Subcommittee on Science and Technology of the Senate Commerce, Science and Transportation Comm.*, 96th Cong. (1979) (statement of Malcolm Wallop, Wyoming senator). Copy supplied by Senator Wallop's office to author in 1983. (Wallop's proposal first appeared in an article, "Opportunities and Imperatives of Ballistic Missile Defense," *Strategic Review*, Fall 1979.)

INDEX